Humanly Intelligence for HR Leaders

A Practical Guide Using AI to Strengthen Talent, Culture, Learning, and Performance

Renee Hensley, M.Ed.

Shelf-Help Press

Book cover by Renee Hensley, Ideogram/Canva

Illustrations by Renee Hensley, Ideogram

First Edition – 2026

ISBN (eBook): 978-1-970646-04-7

Published by: Shelf-Help Press, Kansas City, Missouri USA

https://3karatproductions.com/home

Dedication

To the part of me that always asked, "Is there a faster, smarter way to do this?" From selling Apple IIe, IIc, and Macintosh computers to exploring today's world of AI, this journey was never truly about the technology itself. It has always been about unlocking human potential.

And to my immediate family . . . thank you for your unwavering love, support, patience, and encouragement throughout this journey. It has meant more than you know.

HUMANLY INTELLIGENCE for HR LEADERS MATERIALS

Foreword

Why Organizations Can't Afford to "Figure Out AI on the Side"

by Keli Tuschman, Co-Founder of The Humanly Collective and Director of Consulting Services at Landever.AI

***

Before diving into this book, I invited Keli Tuschman to share her perspective on why organizations can't afford to "figure out AI on the side." Her insights reflect years of experience helping organizations navigate change, leadership, and the human side of transformation.

Why You Need Help Migrating to an AI Workplace

Many organizations are eager, but hesitant, about integrating artificial intelligence into their workflows. It's tempting to think you can manage your AI journey internally by "using the 20-somethings for new ideas," or by letting senior leaders experiment with the latest tool like Copilot. Others insist, "AI can't do what we do."

These attitudes are dangerous. They put small and mid-sized organizations especially at risk of falling hopelessly behind.

Remember Blockbuster? Their reluctance to adapt doomed them when Netflix reinvented the industry. Or think back to the early days of the internet, when companies limited access to a few managers. That didn't last, everyone eventually needed the web, and those who hesitated lost their edge.

Here's the hard truth: asking already-overloaded employees to "figure out AI on the side" doesn't work. AI isn't a hobby or a project for a handful of power users. It's a transformational shift, impacting every aspect of your organization: processes, culture, security, and customer experience. Partial dabbling won't help you compete with companies that invest fully.

We've all seen what happens when new systems are launched without real investment in learning. Staff do the bare minimum because they never understood the potential. Capabilities go unused, value goes untapped, and leadership eventually blames the technology, when in reality, the failure was a lack of training, adoption, and change management.

You can't afford to repeat that mistake with AI. The stakes are too high, and your competitors won't wait for you to catch up.

So, what does a smart AI migration look like?

- **Investment in expertise**: Bring in proven professionals, not just tech experts, but strategic thinkers who understand both your business and AI's potential.

- **Strong knowledge partners**: Skip the parade of consultants selling apps. Choose partners who will challenge your assumptions, identify real opportunities, and design scalable, sustainable solutions.

- **Deep, ongoing training**: Adoption isn't a one-and-done event. True value comes when employees are supported over time as they learn to apply AI in their day-to-day work.

- **Holistic integration**: Subject matter experts, technologists, and business leaders must collaborate to ensure AI solves real problems—not just adds more digital clutter.

Partial solutions and halfhearted efforts rarely pay off. If you want your organization to survive—and thrive—you must think long-term and invest wisely. The ROI will follow, but only if you commit fully.

Don't let your business become the next cautionary tale. With the right help, your shift to an AI-enabled workplace can deliver the competitive advantage, and resilience, your future depends on.

About the Foreword Contributor

Keli Tuschman is the Co-Founder of The Humanly Collective and Director of Consulting Services at Landever.AI. With more than two decades of leadership experience across public, private, and nonprofit sectors, she helps organizations navigate AI adoption, organizational transformation, and workforce change through practical, people-centered strategies.

Preface

Human Firewall

We are living through the most significant disruption to work since the Industrial Revolution.

But for those of us in the people business? It does not feel like a revolution. It feels like a high-speed collision.

On one side: the seductive promise of efficiency engines. AI tools that claim to solve recruiting volume, policy questions, and performance gaps with the click of a button. On the other: the messy, nuanced, deeply human reality of the people we are actually responsible for.

And right now, those two worlds are crashing into each other . . . fast.

Human Resources has entered what I call the **Grey HR Zone**.

The **Grey HR Zone** is where automated logic meets human consequence. It is where a data-driven algorithm decides who gets an interview based on a keyword. Where a system flags someone as a flight risk without ever understanding their story. Where an automated offboarding workflow deletes access for an employee on medical leave, because the system saw a date, but never saw the person.

The system works. Until it doesn't. And when it doesn't, people pay the price.

I have spent decades inside organizations across healthcare, banking, technology, and education. Leading transformation. Making things actually work in environments where time is limited, budgets are tight, and people do not get a second chance to figure it out later.

I have seen what happens when technology is introduced without clarity. I have seen leaders chase tools instead of outcomes. And I have seen what happens when the human side of work is treated as an afterthought.

I wrote this book because the gap is real, and it is getting wider by the day.

We are being asked to implement technologies we do not fully understand, trust vendors who promise bias-free systems that would not hold up under real scrutiny, and lead organizations through change while we ourselves are still sorting out what is real and what is hype.

Meanwhile, the noise is relentless. Everyone has a tool. Everyone has a platform. Everyone has a promise. Very few have accountability.

> Here is the reality most people will not say out loud: if HR does not step into this space . . . fully, confidently, and strategically, someone else will. And they will not be thinking about trust, fairness, belonging, or the human cost of the decisions being made.

This is not another "*AI is magic*" book. You will not find Silicon Valley fairy tales or perfectly packaged case studies that ignore the complexity of real organizations. This is a practical, slightly scrappy, and deeply human guide for the people actually doing the work. The ones in the middle of it. The ones who do not get to experiment without consequences.

Whether you are a solo HR leader in a 200-person company or the CHRO of a global organization, the mission is the same: we have to build a Human Firewall.

The premise of this book is simple but non-negotiable. AI should automate the mundane so that we can be more human. not less. We leverage AI where it belongs. But we fiercely protect the moments that require judgment, empathy, context, and ethical reasoning. Because those are not features. Those are responsibilities.

Most AI guidance tells you what to automate. This book tells you what not to. And it tells you how to tell the difference.

If we get this right, AI will not replace HR. It will finally remove the administrative weight that has kept us stuck. So, that we can step fully into the role we were always meant to play. Strategic. Human-centered. Culture-shaping. Decision-influencing.

If we get this wrong, we do not just risk inefficiency. We risk trust. We risk fairness. We risk becoming a function that manages systems instead of one that leads people.

This is about ownership. This is about responsibility. This is about whether we are willing to step into this moment and say: we will not automate what should be human.

Let's get to work.

Renee Hensley, M.Ed. Co-Founder, The Humanly Collective

Contents

Introduction

When AI Meets the Human Moment

The ballroom buzzed with talk of algorithms, data patterns, and disruptive intelligence. At a center table, an HR executive sat quietly, scanning slides packed with neural networks and predictive analytics. She flipped through her notes, trying to translate each new acronym into her daily reality of recruiting, onboarding, performance reviews, training, culture, etc.

The longer the keynote went on, the wider the gap felt.

As the applause died down, she leaned toward a colleague and whispered:

"But how do we actually use this without losing what makes us human?"

If you have ever had that thought, caught between the rising tide of AI hype and the steady ground of human connection . . . you are not alone.

And you are not behind. You are paying attention.

The Pressure Is Real and So Are the Stakes

Today's People & Culture leaders are navigating something genuinely new. Not just a technology shift, but a values test. Boards demand innovation. Executives chase efficiency. Employees worry about their jobs, their data, and their sense of belonging.

AI is everywhere. So is the skepticism, the confusion, and the quiet fear that sits underneath every vendor presentation that promises to 'transform your HR function from the inside out.'

The stakes go far beyond productivity metrics. They reach into trust, fairness, and the soul of your workplace. And the HR leaders who are going to navigate this well are not the ones who adopt AI fastest. They are the ones who are most deliberate about where human judgment must remain, and most uncompromising about protecting that space.

That is what this book is about. Throughout this book, I use the term people function to describe the full ecosystem of Human Resources, Talent Acquisition, Learning and Development, Organizational Development, Employee Relations, and People Operations. Titles vary across organizations. The responsibility does not.

Why You Can Trust This Perspective

I have spent decades helping organizations navigate moments like this. As co-founder of The Humanly Collective, and a lifelong advocate for workplaces where people genuinely matter, my work spans healthcare, banking, technology, education, retail, sales, and real estate. I have led learning, culture, and organizational transformations through periods of rapid change.

I have seen what happens when organizations rush toward new technology without grounding it in human reality. And I have seen the opposite mistake, when fear or inertia causes leaders to miss meaningful opportunities.

I have also worked inside the technology. I understand what AI can and cannot do, not as an observer from the outside, but as a practitioner who has sat at both tables. That dual perspective is the foundation of everything in these pages.

Here is what experience has taught me: technology evolves quickly. Human behavior does not. Sustainable change lives at the intersection of both.

This book is further grounded by two contributors whose real-world implementation experience informs key frameworks throughout these pages: Alan Landever, Founder and CTO of Landever.AI, a NASA Educator Fellow and AI architect who has spent two decades making complex technology accessible to non-technical leaders; and Keli Tuschman, Director of Consulting Services at Landever.AI, whose career as a CHRO, VP of Consulting, and HR Director brings the practitioner voice that no framework alone can replace.

What Is Actually Breaking Right Now

Before we talk about solutions, we need to be honest about the problem. Not the theoretical future problem, but the things that are going wrong in HR functions right now, in organizations that have already started down the AI path. Chapter 1 names five fracture points that are showing up repeatedly across every size of organization that has begun the AI journey without a governance framework to guide it. They are worth reading with your own people function in mind.

THE HUMANLY INTELLIGENCE MODEL

At the heart of this book is a three-zone framework for every AI decision in people work:

ZONE 1 - AUTOMATE: Workflows AI should own entirely. Administrative, repetitive, low-stakes, reversible.

ZONE 2 - COLLABORATE: Workflows where AI and human judgment work in tandem. AI provides a signal; humans provide a decision.

ZONE 3 - PROTECT: Decisions that must always stay human. Legally sensitive, emotionally complex, irreversible.

Every chapter applies this model to a specific domain of people work.

By the time you finish, you will be able to apply it to any AI decision you face, including ones that don't exist yet.

Why This Book Exists

This book was born from moments of tension: when leaders feel torn between the future and the familiar, between innovation and integrity, between what's possible and what feels safe.

Our purpose is simple but urgent: to help people leaders use AI in ways that amplify what makes us most human, not erase it. To position HR not as a downstream implementer of AI decisions, but as the ethical compass of the digital era.

AI can help us make better decisions. It can surface talent more fairly. It can free people from low-value work so they can focus on what actually matters. But only if we treat it as a tool in service of human judgment, not a replacement for it.

The organizations that succeed with AI in HR don't start by asking: "*What can this technology do*?" They start by asking: "*Where do our people struggle to execute and how can we design better support*?" That distinction changes everything.

Let's Address the Doubts - Head On

You may be wondering:

- Will AI replace jobs or create new inequities?
- Can it actually reduce bias, or just automate it at scale?
- How do we measure ROI without losing sight of people?
- How do we move forward when leaders or teams resist change?
- What happens when an AI tool gets something wrong about one of my people?
- Who is responsible when an algorithm makes a call nobody reviewed?

These concerns are real. We've faced them ourselves. This book doesn't gloss over the risks. It helps you navigate them honestly, with practical steps, clear guardrails, and frameworks grounded in real-world people work.

How This Book Tackles HR Challenges

This book is designed to solve problems, not just explain them. Each chapter opens with a real-world scenario (CASE IN POINT), the kind that keeps you awake and the kind you have likely faced or seen unfold in your own organization.

From there, we move quickly through the framework, case, tools, and guidance for the grey areas, so you leave each chapter with something you can apply immediately.

Here is our promise to you:

- No empty buzzwords
- No technical rabbit holes
- No vague optimism without accountability

Instead, every chapter delivers:

- A real-world opening vignette, a scenario you will recognize immediately
- Plain-language explanations of complex AI concepts
- Practical frameworks you can use the same week
- Cross-industry case studies from organizations of all sizes, not just Silicon Valley
- Ethical playbooks, decision guides, and checklists
- A Grey Area Decision Framework, where human judgment must override AI every time
- A prompt library and workbook exercises for solo use and team facilitation

This is a working guide, not a thought experiment. The conclusion brings it all together with a 90-day action plan, three months of specific, sequenced actions that move you from audit to implementation to leadership.

Not someday. Starting Monday.

What You Will Walk Away With

By the end of this book, you will:

- Know exactly which AI workflows are safe to automate and which carry legal, ethical, or cultural risk if you do.
- Have a human-in-the-loop framework for every agentic process running in your people function.
- Be able to audit any AI tool for algorithmic bias without a data science degree.
- Know how to evaluate vendors beyond the sales pitch]and the red flags that should disqualify a tool immediately.
- Have the language to lead AI adoption conversations with executives, teams, and skeptical stakeholders.
- Be able to measure impact and prove ROI in the language your CFO and CEO respond to.
- Design ethical, people-centered processes where innovation and integrity grow together.

COMPANION RESOURCES

All templates, checklists, and prompt libraries in this book are available as downloadable resources at the Shelf-Help Press | Humanly Intelligence for HR Leaders resource site. Google Drive link: https://dr

HUMANLY INTELLIGENCE for HR LEADERS MATERIALS

A Final Invitation

This is a moment that calls for clarity and courage.

The future of work is not just about technology. It is about defining what intelligence means, together. It is about deciding, deliberately and with full awareness of the stakes, which decisions belong to AI and which decisions belong to the humans whose lives those decisions affect.

You are invited to rethink what it means to be intelligent at work. To become a pioneer of a humanly intelligent organization. To use this book as your blueprint for progress without compromise.

People & Culture leaders are the guardians of trust, fairness, and belonging. AI should serve those values, not undermine them. And the HR leaders who hold that line, who know exactly where the human must remain, and refuse to surrender that ground to efficiency arguments or vendor promises, are the ones who will define what great people work looks like in the years ahead.

The future of HR is not artificial. It is humanly intelligent.

HOW TO USE THIS BOOK
Three reading paths based on your situation:
THE TIME-POOR LEADER: Read the opening vignette, section headers, Grey Area Decision Framework, and Chapter Checklist in each chapter. Essential framework in under 20 minutes per chapter.
THE PRACTITIONER: Read every chapter in sequence. Complete the workbook exercises as you go. Implement at least one checklist item before moving to the next chapter.
THE TEAM LEADER: Use the book as a facilitated learning program. Each chapter contains enough material for a 90-minute team session. The **Skeptic's Corner FAQ** in each chapter is designed for group discussion.
All templates, checklists, and prompt libraries are available as downloadable resources at the companion website.

Worth Remembering

The woman in the ballroom asked the right question. Most organizations are still working up the courage to ask it.

> *74% of workers see AI as a complement to human capabilities, not a replacement for them. Only 11% of organizations have embedded AI into daily workflows. The gap is not technological. It is leadership.* ~ SHRM, From Adoption to Empowerment, 2025

Chapter 1: An Honest Diagnosis of AI in HR

What's Actually Broken

CASE IN POINT

Maya sat in her office, the blue light of her monitor reflecting in the dark window. It was 8:30 PM on a Tuesday. As the Head of People for a 400-person logistics firm, she was used to long hours., but this was different. This wasn't a standard crunch period. It was a crisis of confidence.

Figure 1.1 - Maya's Story

Six months ago, her CEO had mandated the implementation of a cutting-edge AI recruitment platform. The promise was seductive: 70% faster screening, "unbiased" candidate matching, and a significant reduction in the workload for her overstretched talent team. For four months, it seemed to work. The metrics looked stellar. Then came the email from a candidate's lawyer.

The complaint was surgical. It alleged systemic discrimination, claiming the AI had screened out 94% of applicants from a specific postcode associated with a high percentage of ethnic minority residents. Maya scrambled for an explanation, only to find a black box. She had no audit trail. She had no override protocol. She didn't even know the tool was using geographic data as a proxy for "commute stability."

In 48 hours, Maya has to stand before the board. She cannot just tell them the tool failed. She has to explain why she let it happen. And, more importantly, how she will ensure it never happens again.

Maya's story is not an outlier. It is the quiet reality of the AI gold rush in Human Resources. We have been sold a vision of efficiency that often ignores the messy, analogue, and deeply legalistic reality of managing human beings. To fix the future, we must first have the courage to diagnose exactly what is currently broken.

***

Before this chapter offers a single solution, it is going to do something most AI books avoid: **name what is actually broken**. Not what might break. Not what could theoretically go wrong in a future where AI becomes more powerful. What is broken right now, today, in organisations that have already implemented AI in their people functions and are living with the consequences.

This is not a pessimistic chapter. It is a diagnostic one. A doctor who tells you the truth about your symptoms before prescribing treatment is not being unkind. They are being responsible. The same principle applies here. You cannot design the right solution until you have named the right problem.

And the problem is not that AI is too powerful, or that it is being adopted too quickly, or that the technology is inherently dangerous. The problem is simpler and more solvable than any of those things. Most organizations are implementing AI in their people functions without a framework for knowing where human judgment must remain. That gap, between what AI is doing and what a human should be overseeing, is where the risk lives.

This chapter maps that gap. It names the five places where AI is currently failing HR professionals, their organizations, and the people they serve. And then it introduces the framework that the rest of this book is built on: the Humanly Intelligence Model. This is a practical, **three-zone decision architecture** that tells you exactly where AI should work, where AI and humans should work together, and where humans must never step back.

The Five Fracture Points

The following five fracture points were identified through extensive research across HR practitioner communities, book reviews, and field conversations with people leaders across sectors. They are not edge cases. They are not predictions. They are patterns that are showing up repeatedly, in organizations of every size, in every industry that has begun the AI journey without a governance framework to guide it.

Fracture Point 1: The Hallucination Liability

AI language models are designed to produce fluent, confident, well-structured responses. What they are not designed to do is verify that those responses are accurate. This is not a bug that will eventually be fixed. It is a structural feature of how these systems work. They are probability engines, not truth engines that are predicting the next most statistically likely word, sentence, or paragraph, not the most factually correct one.

In most business contexts, this limitation is manageable. A hallucination in a marketing email is a funny quirk. A hallucination in HR is a legal hand grenade. When an employee asks an AI-powered internal tool about maternity leave entitlements, disciplinary procedures, or grievance rights, they expect 100% accuracy. If the AI invents an extra week of paid leave, omits a critical step in a formal process, or misrepresents the organization's policy on reasonable adjustments, the company is liable, regardless of which system produced the guidance.

HR professionals are uniquely exposed to this risk because the questions their people ask are uniquely high-stakes. *"Am I entitled to this?" "What are my rights in this situation?" "What does the policy say about this?"* These are not abstract queries. They are questions with legal, emotional, and relational consequences if answered wrongly.

Practitioners in HR communities have a word for AI-generated content that looks authoritative but requires exhaustive fact-checking before it can be trusted: "**AI slop**." Content that consumes more time to verify than it would have taken to produce correctly from scratch. The burden of fact-checking now falls squarely on the HR practitioner, who often spends more time managing the AI output than they would have spent completing the task themselves.

The **hallucination liability** is not a reason to avoid AI in HR. It is a reason to design every AI-assisted workflow with a **human review gate** between the AI output and the employee it will affect. That design decision is at the heart of the **Humanly Intelligence Model**.

Fracture Point 2: The Empathy Gap

AI is binary. HR is analogue. Algorithms thrive on data points, standardized inputs, and predictable outputs. Human life, however, is lived in the **grey areas**.

Consider a performance conversation. On paper, an employee might be failing to meet their KPIs. An AI can identify that trend and even draft a standardized improvement plan.

But an AI cannot:

- See the flickering hesitation in an employee's eyes that signals a brewing mental health crisis
- Understand the nuance of a bereavement, or the
- Cultural friction within a team that isn't captured in a sentiment analysis dashboard. It cannot read the room

When organizations automate the delivery of sensitive HR functions, such as; grievance acknowledgements, performance warnings, employee relations triage, they lose the human-to-human connection that prevents litigation and builds culture. When empathy is automated, it ceases to be empathy. It becomes a script. And employees know the difference.

This is not an argument against using AI in sensitive HR contexts. It is an argument for being deliberate about where it is used and what role it plays. AI can support an HR professional preparing for a difficult conversation. It cannot have the conversation. That distinction, between AI as preparation tool and AI as decision-maker, is one of the most important lines in this book.

Fracture Point 3: Prompt Burnout

Of all five fracture points, this one receives the least attention in the published literature on AI in HR. It is also, in our experience, the one causing the most damage to the HR profession right now.

Prompt burnout is what happens **when AI adoption is mandated rather than chosen**. There is a massive difference between willing adoption, where a tool solves a felt pain point, and enforced usage, where a tool is handed down by a CEO with a two-hour onboarding session and an expectation that the team will figure it out.

Mid-to-senior HR leaders are being told to "**be more AI-driven**" without being given the time or training to understand the tools they are being asked to champion. The result is a workforce of HR professionals who feel like they are failing at two jobs simultaneously: their actual role in people strategy, and their new, unasked-for role as amateur prompt engineers.

HR leaders are the most at-risk group for this particular burnout because the demand placed on them is structurally contradictory. They are tasked with maintaining human culture while being required to interact with systems that feel, at their worst, fundamentally inhumane. They are spending **significant cognitive load** prompting, reviewing, correcting, and re-prompting. And underneath all of that is the quiet, corrosive suspicion that they are training a system that someone in the leadership team will eventually point to as evidence that the HR function no longer needs as many people.

If you have felt this, you are not alone. And you are not failing. You are experiencing a structural problem that this book is designed to help you solve.

Fracture Point 4: Recruitment Pollution

The recruitment landscape is currently locked in an AI arms race. Candidates are using AI to generate hundreds of tailored CVs and cover letters at the click of a button. In response, HR teams are deploying AI to screen those same applications. We have reached a point where AI is effectively talking to AI, while the human beings on both sides are increasingly alienated from the process.

This creates what we call recruitment pollution which is a flood of high-volume, low-quality applications that masks genuinely strong talent. When we rely solely on AI filters, we prioritize candidates who are best at gaming the algorithm rather than those who are best for the role. The AI optimizes for keyword density and pattern matching. The candidates who write about their experience in their own words, who describe their achievements using human language rather than job description language, are filtered out.

The quality collapse is measurable. HR professionals are reporting declining shortlist quality, rising time-to-fill, and a growing sense that the efficiency gains promised by AI recruitment tools are being consumed by the quality problems those same tools are creating. And the candidate trust crisis is compounding it: the best candidates, the ones with options, are opting out of bot-led processes entirely.

Fracture Point 5: Agentic AI Without Guardrails

Most organizations are still thinking of AI as a chatbot, something you talk to. But we are rapidly moving toward **agentic AI**: systems that don't just draft a response, but actually execute the task.

An agentic AI could, in theory, identify a performance dip, pull the relevant policy, draft a warning letter, and send it to the employee. All this without a human ever clicking approve. This is the fracture point most organizations have not noticed yet. Without strict guardrails, we are handing over the execution of HR to systems that lack judgment. Agentic AI is a powerful engine. But right now, many HR departments are driving it without a steering wheel.

In many organizations, agentic HR workflows are already running. Automated offboarding sequences that revoke system access and initiate payroll changes. Benefits enrollment triggers that process elections without human review. Compliance training escalation workflows that flag employees and notify managers automatically. Some were implemented by HR. Many were implemented by IT or operations AND without HR's knowledge, and without any consideration of where the human checkpoints should be.

When agentic workflows work correctly, they are genuinely valuable. When they do not, the consequences land on real people. This fracture point has its own dedicated chapter later in this book because it deserves one. For now, know this: if you do not know which agentic workflows are currently running in your people function, finding out is your most urgent priority.

THE ICEBERG BELOW YOUR AI ADOPTION

The visible impact of AI in HR, the drafting tools, the screening platforms, the chatbots, represents a fraction of what is already happening.

Research from Project Iceberg identifies two layers of AI transformation:

The Surface (2.2% of wage value): visible tech adoption in roles like software development and data science.

The Hidden Mass (11.7% of wage value, 5x larger): cognitive automation expanding into Finance, HR, Customer Support, and Administration.

HR sits firmly in the hidden mass. The transformation affecting your function is not the one making headlines.

It is the one already running in your HRIS, your payroll system, and your onboarding workflow.

(Source: Project Iceberg, contributed by Alan Landever, Landever.AI)

Figure 1.2 - Iceberg Below Your AI Adoption

The Humanly Intelligence Model

Naming what is broken is only useful if it leads somewhere. The five fracture points in Section 1.1 are not a list of reasons to avoid AI. They are a map of where the governance is missing.

And governance, in this context, does not mean bureaucracy or restriction. It means knowing, for every AI-assisted workflow in your people function, exactly where human judgment must live, and designing your processes around that knowledge.

That is what the Humanly Intelligence Model is for.

The model divides every AI-assisted HR workflow into three zones. Not by function. Not by technology. By what is at stake if the AI gets it wrong, and whether a human needs to be in the decision before, during, or instead of the AI output.

Zone 1 – Automate: AI owns this

Zone 1 workflows are those where AI can own the output entirely. They share four characteristics: they are low stakes (a wrong output does not harm an individual), they are administrative (they process information rather than make judgments about people), they are repetitive (the task is performed frequently and consistently), and they are reversible (if the AI gets it wrong, the error can be caught and corrected without lasting consequences).

The test: if the AI's output were completely wrong, what would happen? If the answer is "*someone would need to rewrite a document*" or "*we would resend the communication*," Zone 1 is appropriate. Job description drafting. FAQ responses to policy queries. Interview scheduling. Training reminders. These belong in Zone 1, with a light human review gate and a reversibility mechanism.

Zone 2 – Collaborate: AI and human in tandem

Zone 2 workflows are those where AI adds genuine value as a signal generator, but where the final decision must rest with a human. AI surfaces the information, identifies the pattern, or generates the options. A human evaluates and decides. The distinction between Zone 2 and Zone 1 is not the complexity of the task, it is the stakes attached to the output.

Candidate shortlisting. Learning and development recommendations. Engagement survey analysis. Skills gap identification. Performance data aggregation. These are Zone 2 workflows. AI is doing valuable work. But a human is reviewing that work before it influences any individual decision. That review is not optional. It is the governance mechanism that makes Zone 2 safe.

Zone 3 – Protect: Human owns this. Always.

Zone 3 workflows are those where human judgment must own the decision entirely. AI may provide information and context. It must never make, or appear to make, the call.
Zone 3 decisions are legally sensitive, emotionally complex, and irreversible. Getting them wrong does not produce a document that needs rewriting. It produces a harmed employee, a legal claim, or a trust breakdown that takes years to repair.

Grievances. Disciplinary processes. Terminations. Mental health disclosures. Reasonable adjustments. Pay equity decisions. Performance improvement outcomes. These are Zone 3. Always. Without exception. Think of Zone 3 as the human firewall around the most sensitive parts of the employee experience. The goal is not efficiency. The goal is protecting the people and the organization from the consequences of removing human judgment from decisions that require it most.

Regardless of how much AI could theoretically handle, regardless of how under-resourced your team is. The resource pressure argument for automating Zone 3 is the most dangerous argument in this book, and Chapter 3 addresses it directly.

THE CENTRAL DECISION TEST

When you are unsure which zone a workflow belongs in, apply this single question:

"Would the person affected by this decision expect a human to have made it?"

If yes — Zone 3. Always.
If maybe — Zone 2, with a human checkpoint before any output reaches the individual.
If no — Zone 1, with a review gate and a reversibility mechanism.

If you are asking the question at all, the workflow is probably Zone 2 or Zone 3.
Uncertainty always resolves in favor of human judgment

Figure 1.3 - Decision Test

Most organizations are currently operating with no model at all. Not because their leaders are careless, but because the tools arrived faster than the governance framework. AI platforms were procured, implemented, and handed to HR teams with onboarding sessions and vendor promises. Nobody stopped to map the workflows. Nobody asked where the human checkpoints were. Nobody designed for the grey areas.

The Humanly Intelligence Model does not require a technology audit, a data science team, or a six-figure consulting engagement. It requires a half-day with your HR leadership team, the Zone Sorter you will find at the end of this chapter, and the willingness to name, honestly, which of your current workflows are operating without adequate human oversight.

The Cost of Getting This Wrong

The five fracture points are not abstract risks. They have concrete, measurable costs. And the organizations that have experienced them are, for the most part, not willing to be named publicly. What follows is drawn from anonymized accounts, published case studies, and the pattern of what goes wrong when AI in HR operates without governance.

Legal and Compliance Exposure

The hallucination liability fracture point is not a reputational risk. It is a legal one.

When AI-generated policy guidance reaches an employee, and that guidance is wrong, the organization does not get to blame the tool. The organization is the employer. The employer is responsible for the guidance its systems provide.

Employment law claims citing incorrect AI-generated advice are already beginning to surface across U.S. jurisdictions. They are not yet common. They will become common as AI adoption scales and the human review layer between AI output and employee impact remains under-resourced, or absent.

The exposure extends beyond individual claims. As AI becomes embedded in hiring, performance management, and employee relations, organizations are creating systems that directly influence employment decisions, often without clear audit trails, oversight protocols, or accountability structures.

That is not just a process gap.

It is a liability.

> **Note**: Organizations operating globally should be aware that regulatory frameworks such as the EU AI Act introduce additional requirements around transparency, documentation, and human oversight for AI systems used in employment decisions. While this book focuses primarily on U.S.-based organizations, global compliance considerations may apply depending on your workforce footprint.

Reputational Damage: Candidate and Employee Trust

Your employer brand is built on how you treat people. If candidates feel like a cold, biased machine is processing them, they will talk. Glassdoor reviews and social media posts can dismantle a decade of employer branding in a single afternoon. Once an organization is labeled as anti-human, it becomes extraordinarily difficult to attract the kind of talent that chooses its employer on values, not just compensation.

Maya's situation illustrates something that is difficult to quantify and easy to underestimate: the reputational cost of AI gone wrong in a people context is disproportionate to the technical error that caused it. Candidates who discover they were screened out by an algorithm operating on flawed parameters do not simply move on. They share their experience in the same online communities where your next strong candidate is deciding whether to apply.

Employees who discover that a sensitive HR interaction was handled by an AI tool, without their knowledge or consent, experience that discovery as a betrayal. Not as a process failure. As a betrayal. Trust, once broken in that way, is not rebuilt by a town hall or a policy update.

The Hidden Cost: HR Credibility with the C-Suite

HR has spent the last two decades fighting for a seat at the strategic table. That seat is predicated on HR's ability to manage risk and people strategy effectively. If HR leaders are seen as being asleep at the wheel while AI creates chaos, that hard-won credibility evaporates. The CEO does not want to hear that the vendor promised the tool was safe. They want to know why HR did not have a framework to verify it.

When an AI tool procured or championed by HR produces a discrimination complaint, a compliance breach, or a candidate experience crisis, the conversation in the boardroom does not focus on the tool. It focuses on the function that endorsed it. The HR leaders who are building governance frameworks now are not being cautious. They are being strategic and they are doing it before the crisis forces the conversation.

The Opportunity Cost: What Getting It Right Looks Like

It would be incomplete to end this section without naming what is on the other side of the governance gap. The organizations using AI well in their people functions are not the ones with the largest technology budgets or the most sophisticated data science teams. They are the ones that have been **most deliberate about where human judgment lives** inside their AI-assisted workflows.

They are hiring faster, with better shortlists. Their HR teams are spending less time on Zone 1 administrative tasks and more time on the complex, high-value Zone 3 work that creates genuine organizational value. They aren't just surviving AI, they are using it to become more human, not less. Their employees report higher trust in HR processes because those processes are transparent about what AI does and what humans decide.

That is where this book is taking you. Not to a future where AI has solved HR. To a present where you have designed your people function to use AI well, with human judgment exactly where it needs to be, and the confidence to prove it.

Cross-Industry Case Study: U.S. Healthcare System

CROSS-INDUSTRY CASE STUDY

U.S. Healthcare System - Highly Regulated Environment
Organization: Regional hospital network, 6,000 employees
Function: HR and Employee Relations

The Problem:

Following a system-wide digital transformation, the organization implemented an AI-assisted triage tool to support employee relations. The tool categorized incoming cases, suggested policy guidance, and routed inquiries to HR business partners. It was deployed quickly, with limited HR involvement in the design and oversight model.

Within three months, a pattern emerged.

Employees submitting formal grievances began receiving AI-generated policy summaries that contained **material inaccuracies**. In multiple cases, the tool referenced outdated procedures that had already been replaced. In one instance, an employee was incorrectly advised about their right to representation during a disciplinary meeting.

The HR team did not identify the issue.

Outside counsel did—after an employee escalated the case.

This was not a healthcare-specific failure. Any organization using AI to interpret policy without human validation is exposed to the same risk.

What They Did:

The HR Director paused the AI triage function and initiated a structured review.

Rather than removing the tool entirely, the team redesigned the operating model around **human oversight and risk segmentation**, implementing a three-tier review protocol:

- **Tier 1:** All AI-generated policy guidance is reviewed by an HR advisor before being shared with employees.
- **Tier 2:** Cases involving protected classes, formal grievances, or disciplinary action escalated for senior HR review.
- **Tier 3:** Mental health disclosures removed entirely from AI triage and routed directly to a human HR business partner, without exception.

Outcome:

- 40% reduction in policy errors within the first quarter of the new protocol
- The HR team's confidence in the tooling significantly recovered
- Zero mental health disclosure cases processed by AI since Tier 3 was implemented

The Grey Area This Case Exposes

Mental health disclosures are not a process category. They are a **human responsibility**.

No efficiency argument, no resource constraint, and no volume pressure justify automation in this space. The moment an employee discloses a mental health concern, the requirement is not speed or scale.

It is **human connection, judgment, and care**.

Figure 1.4 - Cross-Industry Case Study

Chapter Checklist

☐ I have listed every active AI tool currently used across HR, Talent, OD, and L&D, including those implemented by IT or vendors without HR input

☐ I have **identified at least one workflow** that carries legal or ethical risk if AI output is wrong.

☐ I have **mapped at least three workflows to Zone 1, Zone 2, or Zone 3** using the Humanly Intelligence Model.
☐ I have **identified the first Zone 3 workflow** running without adequate human oversight. This is my immediate priority.
☐ I have **shared the five fracture points with at least one colleague** and discussed which ones are most urgent for my organization.

Prompt Library & Workbook Prompts

DIAGNOSTIC PROMPT

[List every HR workflow in [organization name] that currently uses AI in any form. For each workflow, describe: what the AI does, who reviews the output before it reaches an employee, and what happens if the output is wrong. Identify any workflow where there is currently no human review gate.]

ZONE SORTER PROMPT

[I am going to describe an HR workflow. Tell me whether it belongs in Zone 1 (safe to automate), Zone 2 (human-AI collaboration), or Zone 3 (human decision only) using the Humanly Intelligence Model. Explain your reasoning and identify the specific risk that places it in that zone.]

WORKBOOK EXERCISE

Write down the one AI workflow in your organization that worries you most. Be specific: what does the AI do, who does it affect, and what is the worst realistic outcome if it gets something wrong? What would need to be true for you to feel confident about it? That answer is your governance requirement.

TEAM EXERCISE

Share the five fracture points with your HR leadership team. Ask each person to rank them in order of urgency for your organization, privately, before the group discussion. Then compare rankings. The disagreements in that conversation are more valuable than the agreement. They tell you where your team's mental models of AI risk diverge, and where your governance framework needs to be most explicit.

SKEPTIC'S CORNER FAQ

Q: We've been using AI tools for a year with no problems, isn't this overblown?

A: The absence of a visible problem is not the same as the absence of risk. Research consistently shows that algorithmic bias and AI liability issues surface six to eighteen months after deployment, often when a demographic pattern in the data becomes visible, or when a specific affected individual decides to escalate. The question is not whether something has gone wrong yet. It is whether you would know if it had. If you do not have an audit trail, a bias monitoring process, or a human checkpoint on your AI-assisted decisions, the honest answer is: you would not know.

Q: Our vendor assured us their tool is bias-free.

A: No tool is bias-free. Any vendor who claims otherwise is either mistaken about how their system works or is using marketing language that should disqualify them from your consideration. Bias is not a design choice that vendors make. It is a structural property of machine learning systems trained on historical data. The question is not whether bias exists in your tools, it does. The question is whether you have a process to detect it, measure it, and respond to it. Chapter 5 gives you that process in full.

Q: We don't have the budget for a full AI audit.

A: The **Grey Area Decision Framework** in this chapter costs nothing but time. A half-day workshop with your HR team using the **Zone Sorter** will give you a clearer picture of your governance gaps than a six-figure vendor review, because it is grounded in your specific workflows, your specific tools, and your specific risk exposure. Start there. The audit comes later, after you know what you are auditing.

What This Chapter Has Asked You to Do

It has asked you to look at your current AI landscape honestly. Not with the optimism of the vendor presentation, and not with the anxiety of the worst-case scenario. With the clear-eyed pragmatism of an HR leader who knows that the people in their organization are affected by every decision this technology makes. And, who is not willing to outsource their judgment about where the humans must remain.

The five fracture points are not failures. They are signals. Each one is pointing to a governance gap that the **Humanly Intelligence Model** is designed to close. And each chapter that follows takes one dimension of that model and shows you, practically and specifically, how to apply it in the context of your own people function.

Maya's board meeting is in 48 hours. She cannot go back and redesign the procurement process.

But she can walk into that room with something more valuable than an apology: clarity. Clarity on what failed. A framework to prevent it from happening again.

And the language to explain, clearly and confidently, what it means to lead AI adoption with human judgment at the center.

That is Humanly Intelligence.

> **What's Actually Broken** – *Only 1% of layoffs in the first half of 2025 resulted from AI increasing employee productivity. Yet, organizations are being asked to cut headcount on the basis of returns that have not yet materialized.*
>
> ~ Gartner, 2026

Chapter 2: Know Your AI

What It Can and Cannot Do in People Work

CASE IN POINT

James is a VP of Talent Acquisition at a global retail brand with over 15,000 employees. He is a seasoned professional. He has navigated global recessions, massive restructuring, and the Great Resignation with poise. But lately, he has felt like a stranger in his own profession.

Figure 2.1 - James' Story

In the last three months, James has sat through four AI vendor demonstrations. Each presenter arrived with a polished deck and the same set of buzzwords. One promised their tool would eliminate bias once and for all through a proprietary neural network. Another claimed their algorithm could predict flight risk with 98% accuracy, allowing James to intervene before a high-performer even knew they wanted to leave.

Throughout these meetings, James nodded. He smiled when they mentioned vector embeddings and parameter weights. He did not want to look like he was behind the curve. He certainly did not want to admit that he had no idea how a machine could measure cultural fit better than his senior recruiters. Eventually, under pressure from the Board to modernize the stack, he signed a contract for a comprehensive AI talent suite.

Six months later, the tool is live, and James is paralyzed. The bias-free screening is filtering out candidates with non-traditional backgrounds that his team used to love. The flight risk dashboard is flagging employees who just celebrated their five-year anniversaries, and he has no idea why. He bought the dream. He does not understand the machinery.

***

This chapter is for James and for every HR leader who has felt the pressure to adopt AI without a clear map of what is actually happening under the hood. You do not need to learn Python. But you must move from nodding along to asking the right questions.

You do not need to become a data scientist to lead AI well in your people function. You do not need to understand neural networks, write a line of code, or pass a machine learning certification. What you do need is something more fundamental and more immediately useful: an honest, plain-language understanding of what AI actually is, what it genuinely cannot do, and how to ask the right questions of the people who are trying to sell it to you.

That is what this chapter provides. It is the technical foundation written for non-technical people. By the time you finish it, you will be able to hold your own in any AI conversation; with vendors, with your IT team, or with your CEO. Not because you have memorized the jargon, but because you understand the substance underneath it.

And that understanding will change how you operate in the **Grey HR Zone**. Because most of the governance failures we explored in Chapter 1, such as the hallucination liabilities, the bias exposures, and the agentic workflows running without oversight, are not failures of technology. They are failures of understanding. James did not fail because he bought a bad tool. He failed because he did not know enough to evaluate what he was buying.

The ground has shifted. Analyst *Josh Bersin* has been tracking the forces reshaping work for decades, and his conclusion is unambiguous: the talent shortages HR leaders are managing today are not a post-pandemic hangover. They are structural. We have entered what ***Bersin calls the post-industrial age***. This is an era defined by continuous workforce scarcity, skills-based hiring replacing credential-based hiring, and AI accelerating the productivity of leaner, more agile organizations.

AI Demystified – The Non-Technical Primer

Let's start with the most important sentence in this chapter, borrowed from Alan Landever, Founder & CTO of Landever.AI and NASA Educator Fellow:

> "Deep down, an LLM is just math. It is not thinking. It is not conscious." – Alan Landever

That statement cuts through more confusion than any technical definition ever could.

Because most of the anxiety surrounding AI in HR is rooted in a misunderstanding: we talk about these systems as if they are making decisions, forming opinions, or exercising judgment.

They are not.

What AI Is Actually Doing

As Alan Landever explains:

The brain behind these systems, the **Large Language Model**, is a massive statistical engine. It predicts the next most likely word in a sequence based on patterns it has learned from vast amounts of data.

It is not retrieving verified facts. It is not reasoning from first principles.

It is doing math.

Highly sophisticated math, at enormous scale, but math nonetheless.

What Large Language Models Actually Do

A large language model (**LLM**) is trained on vast quantities of text; books, articles, websites, research papers, code, conversation transcripts. During training, the model learns statistical relationships between words, phrases, concepts, and contexts. It develops an extraordinarily sophisticated sense of what typically follows what.

Think of the autocomplete feature on your smartphone. When you type "I'll see you..." it suggests "later" or "there." An LLM does this on a massive, planetary scale. It has read billions of pages of text and learned the mathematical relationship between words. When you ask it to write a job description for a Marketing Manager, it is not thinking about what a marketer does. It is calculating the probability of which words usually follow each other in the context of marketing job descriptions it has encountered in its training data.

When you prompt an LLM, it generates a response by predicting, token by token, what text is most likely to follow your input. The result is often impressively accurate. It is also structurally incapable of being certain it is correct, because it is pattern-matching, not fact-checking.

This is why "intelligence" is a misleading word for what these systems do. Intelligence implies understanding. What LLMs do is sophisticated pattern recognition at scale. The distinction matters enormously in HR, where the difference between a plausible-sounding answer and a correct one can be the difference between good guidance and a legal claim.

Generative AI vs. Agentic AI: The Distinction That Defines Your Risk

Most HR leaders are familiar with **generative AI**, tools that produce content in response to a prompt. You ask, it generates. A draft job description. A policy summary. A set of interview questions. A performance review template. You receive an output. A human reviews it. The human decides what to do with it.

Agentic AI is a fundamentally different proposition. Agentic AI does not produce outputs for review. It executes actions. It is given a goal, such as; complete this onboarding workflow, process this benefits enrollment, escalate this compliance training non-completion and it takes the steps required to achieve it, autonomously, without requiring a human to approve each action.

THE GENERATIVE VS. AGENTIC DISTINCTION

GENERATIVE AI: You ask → AI produces output → Human reviews → Human decides
Example: AI drafts a job description. You edit it. You post it.

AGENTIC AI: AI is given a goal → AI executes the steps → Outcome happens
Example: AI detects an onboarding task is overdue → sends a reminder → escalates to the manager →**AGENTIC AI:** updates HRIS — no human in the loop.

The **Grey HR Zone** expands dramatically with agentic AI.
Because every step the agent takes autonomously is a step where human judgment has been removed.

Figure 2.2 - The Generative vs. Agentic Distinction

This distinction matters more in HR than in any other business function precisely because HR decisions are the ones most likely to affect an individual's employment, well-being, and livelihood. A generative AI that produces an incorrect contract clause can be caught by a human reviewer. An agentic AI that executes an incorrect employment status change cannot be caught until after the damage is done.

What AI Cannot Do – The Permanent Limitations

This is not a list of things AI cannot do yet. It is a list of things AI cannot do structurally, limitations that are features of how these systems work, not bugs that will be patched in the next release.

- AI can process the information you give it. It cannot access the history, relationships, and organizational context that an experienced HR professional carries. It does not know that the manager filing a performance complaint has a history of targeting specific demographic groups. It does not know that the employee responding with apparent disengagement has just experienced a bereavement that was not formally disclosed. Contextual judgment.

- AI can analyze sentiment in text. It cannot read the room. It cannot detect the hesitation in someone's voice that signals a mental health crisis beneath a surface-level query about leave entitlements. Sentiment analysis and emotional intelligence are not the same thing. Emotional intelligence.

- AI can apply rules. It cannot reason about competing ethical obligations, weigh the interests of different stakeholders, or navigate the moral complexity of a genuine grey area. It will produce an answer that pattern-matches to ethical language. That is not the same as ethical judgment. Ethical reasoning.

- AI can summarize legislation. It cannot interpret how legislation applies to a specific set of facts, predict how a tribunal would view a specific decision, or give advice that a qualified employment lawyer would stand behind. When an AI sounds certain about a legal question, that certainty is a function of its training data, not of legal expertise. Legal interpretation.

The Hallucination Problem – Plain English

You have already encountered the hallucination liability in Chapter 1. Here is the plain-English explanation of why it happens and why it is permanent.

When an LLM is asked a question, it does not have a strong training signal for it; it does not say "I don't know." It generates the most statistically plausible response. A response that looks and reads like correct information, formatted with the confidence of an expert, but grounded in pattern-matching rather than fact.

In HR, this manifests as:

- Policy citations that are subtly wrong.
- Legislation references that are superseded.
- Process descriptions that omit critical steps.
- Entitlement calculations that are off by a meaningful margin.

None of these errors are obvious in the output. They read fluently. They sound authoritative. They will only be caught by a human who knows enough to check and in many organizations, the human review gate between AI output and employee impact simply does not exist.

The **hallucination problem is not going away**. It is a structural property of **probabilistic language models**. Plan for it permanently.

Design every AI-assisted HR workflow on the assumption that some percentage of outputs will be wrong in ways that are not obviously wrong. The governance response is not to distrust AI. It is to build the human review gates that catch the errors before they reach your people.

The HR AI Landscape – What's Available and What It Actually Does

The market for AI in HR is crowded, fast-moving, and full of overlapping claims. This section maps the major categories of AI tools available to people professionals, what they genuinely do well, and where the specific risks in the **Grey HR Zone** lie for each one.

Recruitment and Sourcing Tools

This is the most mature category of AI in HR and, not coincidentally, the one with the most documented failures.

AI recruitment tools typically perform some combination of candidate sourcing (identifying potential candidates from databases or social platforms), application screening (ranking or filtering applications based on defined criteria), and assessment (evaluating candidate responses to structured questions).

Where they genuinely help: reducing the administrative burden of high-volume screening, increasing the consistency of initial assessment criteria, and surfacing candidates who might be overlooked in manual review.

Where the **Grey HR Zone** appears: any tool that makes or influences a decision about which candidate a human sees is operating in Zone 2 at a minimum. The bias risk is highest here because historical hiring data, the training signal for many of these tools, reflects historical hiring decisions, which frequently reflect historical biases. A tool trained to identify candidates who resemble previous successful hires will perpetuate whatever patterns, conscious or unconscious, shaped those hires.

Employee Relations and Policy Tools - The Hallucination Risk Zone

AI-powered policy assistants, HR chatbots, and employee relations triage tools are rapidly proliferating. The promise is significant: employees get instant answers to policy questions, HR teams reduce the volume of repetitive queries, and case triage becomes more consistent.

The risk is equally significant, and this category sits in the highest-risk part of the Grey HR Zone for the hallucination liability. Policy and legislation change. An AI tool trained on last year's data will answer questions about this year's entitlements with last year's accuracy. Employment law varies by jurisdiction, industry, and collective agreement. An AI tool that cannot distinguish between a general rule and a specific exception is a liability waiting to be activated.

The governance minimum for this category: every output that includes a policy reference, a legislative citation, or an entitlement calculation must be reviewed by a qualified HR professional before it reaches the employee. No exceptions. If the volume makes that impossible, the tool is being used at the wrong scale for your current governance capacity.

Learning and Development Personalization – Where AI Genuinely Excels

L&D is arguably the HR function where AI's genuine strengths align most naturally with the work. Personalized learning pathway generation, content curation from large libraries, skills gap identification, and learning completion tracking are all tasks where AI adds real value and the risk of irreversible harm is relatively low.

The **Grey HR Zone** in L&D is narrower, but it still exists. Any AI that infers skills, potential, or development needs from indirect signals such as communication patterns, meeting participation, or email sentiment has crossed from learning support into surveillance. The transparency test applies. Would employees be comfortable knowing exactly what data is being used about them? If the answer is uncertain, it is no.

Performance Management and Sentiment Analysis – The Privacy and Trust Minefield

Performance management AI and employee sentiment analysis tools represent two of the most ethically contested applications of AI in HR. Both promise valuable insights. Both carry significant risks that are not always visible in the vendor demo.

Performance management AI that aggregates data, identifies trends, and surfaces patterns for manager review can genuinely improve the quality and consistency of performance conversations. Performance management AI that generates ratings, makes promotion recommendations, or triggers performance improvement processes without meaningful human review has moved into Zone 3, where human judgment is non-negotiable.

Sentiment analysis tools that help HR leaders understand aggregate employee mood and surface early warning signals for disengagement serve a legitimate purpose. Sentiment analysis tools that monitor individual employee communications without their knowledge or consent, or that use inferred emotional state to influence employment decisions, are in legally and ethically contested territory in most jurisdictions.

OD and Workforce Planning Tools - The Data Quality Dependency

Organizational design and workforce planning are areas where AI's analytical capabilities offer real strategic value: scenario modelling for future skills needs, organizational network analysis to map how work actually gets done, change readiness assessment, and succession planning support.

The critical dependency for all of these applications is data quality. AI workforce planning tools are only as good as the data they are given. In most organizations, that data is imperfect: job descriptions that do not reflect actual work, performance ratings that reflect manager bias as much as employee capability, and skills data that has never been systematically collected. Before implementing AI in this space, an honest data quality audit is not optional, it is the prerequisite.

Agentic HR Tools - The Category That Changes Everything

Offboarding automation. Benefits enrollment workflows. Onboarding task sequencing. Compliance training tracking and escalation. These are the agentic HR tools that are already running in many organizations. And, often implemented by IT or operations, without HR's input into where the human checkpoints should be.

Chapter 4 is dedicated entirely to agentic AI in HR. For now, the single most important thing to understand about this category: every action an agentic system takes autonomously is a moment where human judgment has been removed from the process. Map every agentic workflow. Identify every moment where an error could reach an employee without human review. Those moments are your Grey HR Zone.

The Vendor Evaluation Framework

James bought a tool he did not understand because he did not know what questions to ask. **This section gives you those questions**. Not the polite questions vendors expect and have prepared answers for. The penetrating questions that separate vendors who genuinely understand their tools from vendors who have memorized a reassuring script.

The **five questions** below should be asked of every AI vendor before any procurement decision. The **Vendor Vetting Matrix** that follows gives you the acceptable answers and the red flags for each one.

The Five Questions

1. How does your model handle data it has not seen before, and what happens when confidence is low? A responsible vendor will have a threshold below which the tool defers to a human rather than generating a low-confidence answer. Any vendor who tells you the tool always provides an answer, with no confidence threshold and no human escalation path, has told you everything you need to know about their approach to hallucination risk.

2. What is your bias testing methodology, who conducted it, and when was it last run? The answer must include a named methodology (the four-fifths rule, disparate impact analysis, or equivalent), a named third party who conducted the audit, the protected characteristics that were tested, and a date. "*We take bias very seriously*" is not an answer. It is a deflection.

3. What does the human override process look like, and how is every override logged? Every AI tool used in HR decisions must have a clear, accessible, documented override path. If overriding the AI requires an IT ticket, a vendor support request, or "*something we can set up for you*," it is not a real override process.

4. What data are you training on, and do we own our outputs? This question has two components. **First**: is your organization's data being used to train or improve the vendor's model? If yes, what are the consent and privacy implications for your employees? **Second**: do the outputs the tool generates belong to you, or does the vendor retain any rights over them? Both questions have legal implications that should involve your legal team before contract signature.

5. What happens when your tool produces a wrong output that reaches an employee, what is the remediation process? A responsible vendor will have a named process, a service level agreement for correction, and a clear contractual position on liability. A vendor who tells you they are not responsible for how their outputs are used has told you that you carry all the risk. That is not a partnership. It is a product sale with a liability waiver.

The Vendor Vetting Matrix

Use this matrix in every vendor evaluation. The green column tells you what a responsible answer looks like. The red column tells you when to walk away.

Area	The Question	Acceptable Answer	Red Flag Answer
Hallucination / Accuracy	How does your tool handle queries outside its training data?	We have a confidence threshold — below X%, the tool defers to a human.	"Our tool always provides an answer." This is an immediate red flag.
Bias Testing	What is your bias testing methodology, who conducted it, and when was it last run?	Named third-party auditor, specific protected characteristics tested, dated report available.	Cannot name the auditor or says, "the model is inherently unbiased."
Human Override	What does the human override process look like, and how is it logged?	Clear escalation path, every override documented, audit trail available to HR.	No override process, or override requires IT involvement to action.
Data Ownership	What data are you training on, and do we own our outputs?	Our data is not used for model training. Outputs are our intellectual property.	Ambiguous language about "improving the service" using client data.
Remediation	What happens when your tool produces the wrong output that reaches an employee?	Named remediation process, SLA for correction, and liability framework in the contract.	"We're not responsible for how the output is used."

Figure 2.3 - Vendor Vetting Matrix

This matrix is available as a downloadable template in Appendix B. Share it with your procurement team, your legal counsel, and your IT partners before any AI HR tool evaluation begins.

Red Flags: When to Walk Away Immediately

- The vendor uses the phrase 'black box' or says the AI's logic is 'too complex to explain.' If the vendor cannot explain what the tool is doing, you cannot govern it.
- The vendor cannot name the third party that conducted their bias audit.
- The contract contains language about using client data to "improve the service" without explicit opt-out.
- The human override process requires vendor involvement to action.
- The vendor claims the tool is "bias-free" or "hallucination-free."
- The liability clause places all responsibility for output accuracy on the client.
- The vendor cannot provide a current, dated audit report on request.
- The demo uses only curated, best-case examples and declines to show edge cases or failure modes.

THE VENDOR CONVERSATION JAMES NEVER HAD

If James had asked the five questions before purchasing, here is what he would have learned:

Question 2 (Bias testing): The vendor had conducted internal bias testing using generalized demographic datasets. James's organization served a materially different talent pool, and no validation had been performed against his actual applicant population.

Question 3 (Human override): The tool had no documented override process. Screening decisions could only be challenged through a vendor support ticket with a five-business-day SLA.

Question 4 (Data ownership): The contract included a clause permitting the vendor to use anonymized interaction data for model improvement. This had not been disclosed to candidates.

None of this would have been visible in the demo.

All of it was in the contract.

The five questions would have surfaced it.

Figure 2.4

Cross-Industry Case Study: U.S. Higher Education

CROSS-INDUSTRY CASE STUDY

U.S. Higher Education - Lean HR Team
Organization: Mid-size university, 1,400 employees
Function: HR policy and employee self-service

The Problem:

The university's People & Culture team, with four people managing HR for 1,400 staff, implemented an AI policy assistant to reduce the volume of repetitive policy queries reaching their inbox. Within three months, the tool began producing incorrect guidance. In two cases, it cited legislation that had been superseded by updated Fair Work Act provisions. In one case, it described a redundancy entitlement calculation using an outdated formula.

The team discovered the errors not through their own monitoring but because an employee's union representative flagged a discrepancy. The team had no process for reviewing AI output before it reached staff.

What They Did:

Rather than decommissioning the tool, the team designed what they called a 'legislation lock', a manually maintained register of current legislation and policy references that all AI outputs were checked against before distribution. The process took one team member approximately 90 minutes per week to maintain.

They also implemented a two-tier review: routine FAQ responses (leave balances, process queries) received a light-touch review by any team member. Any output involving entitlements, legislative rights, or formal processes was reviewed by the most senior HR professional on the team before distribution.

Outcome:

Zero incorrect policy outputs in the 12 months following implementation.

The team built significant internal AI literacy as a by-product of the review process. By building the lock, they learned exactly how the AI read documents, transforming themselves from passive users into confident governors of the technology

The 90-minute weekly maintenance investment was recovered many times over in query volume reduction

The Lesson:

Small teams can build robust AI guardrails without large budgets, data science expertise, or vendor involvement. The 'legislation lock' is a piece of paper and a process, not a technology. The governance solution was human, not technical.

Figure 2.5 - Cross-Industry Case Study

GREY AREA DECISION FRAMEWORK

Grey Area: AI-Generated Policy Advice

This is one of the most consequential grey areas in the HR AI landscape, and one of the most frequently mismanaged.

Use this framework to establish clear boundaries before your next AI policy tool goes live.

NEVER safe to share without senior human review:

- Employment law interpretation
- Reasonable adjustment decisions
- Disciplinary or grievance process guidance
- Redundancy entitlement calculations
- Any output that references specific legislation or case law

SOMETIMES safe (with qualified HR review before distribution):

- General policy summaries
- FAQ responses on process (not entitlement)
- Signposting to the correct policy document
- Process guidance that does not involve legal rights

ALWAYS safe (no review required):

- Directing an employee to the correct HR contact
- Providing the link to a reviewed, current policy document
- Answering non-legal process queries (e.g., how to book leave in the system)

The Tribunal Test:

- If the output could be cited in a tribunal, employment claim, or formal complaint, a human must review it first. No exceptions.
- If you are unsure which category an output falls into, apply the rule: it is in the NEVER category until proven otherwise.

What this means for the Grey HR Zone:

Every AI policy tool you currently use sits somewhere on this spectrum. Mapping where it sits and whether your current review process matches the risk level is one of the most important governance acts you can perform this week. It costs nothing. It takes an hour. And it tells you exactly which of your workflows need a human checkpoint before the next employee query comes in.

CHAPTER CHECKLIST

☐ I can explain in plain English what generative AI does versus what agentic AI does and why the distinction matters for HR governance

☐ I have identified which category (**generative** or **agentic**) each of our current AI tools falls into

☐ I have completed the Vendor Vetting Matrix for at least one current AI tool, using the actual contract, not the vendor's marketing materials

☐ I have identified which of our AI tools carries the highest hallucination risk based on the type of queries it handles

☐ I have a documented process for reviewing AI-generated policy output before it reaches employees

PROMPT LIBRARY & WORKBOOK EXERCISES

EXPLAINER PROMPT

[Explain how a large language model works in plain English, using an analogy relevant to HR professionals. Do not use technical jargon. The explanation should help an HR leader understand both why these tools are valuable and why they sometimes produce incorrect outputs with apparent confidence.]

VENDOR PREP PROMPT

[I am about to evaluate an AI tool for [recruitment / L&D / employee relations / performance management]. Generate 10 penetrating questions I should ask the vendor about: bias testing methodology, data ownership, human override process, hallucination risk, and remediation when the tool produces an incorrect output. Frame the questions so that a evasive or inadequate answer is immediately obvious.]

WORKBOOK EXERCISE

Draw two columns on a page. In column one, list every AI tool your organization currently uses in people work. Be sure to include tools implemented by IT or vendors, not just those HR procured directly. In column two, write one sentence describing what the tool actually does and not simply what the vendor says it does, but what you have observed it doing.

Note: If you find yourself writing "*it just knows*" or "*it's a bit like magic*" then you have identified a knowledge gap. That gap is where your risk is highest. Those are your investigation priorities before the next chapter.

TEAM EXERCISE

Run a vendor pitch roleplay with your HR leadership team. One person plays the AI vendor making a pitch for a new recruitment screening tool. The rest of the team plays the HR evaluation panel using the **Vendor Vetting Matrix**. After the roleplay, debrief on two questions:

- Which of the five questions was hardest to ask?
- Which answer would have been hardest to push back on?

Those answers tell you where your team's vendor evaluation muscle needs strengthening.

SKEPTIC'S CORNER FAQ

Q: "Our IT team manages AI procurement – why do HR leaders need to understand this?"

A: IT manages the infrastructure. HR manages the impact. If an AI tool makes a biased hiring decision, the regulator will not call your CTO, they will call you. Technical ownership does not transfer professional accountability. You would not let IT write your sexual harassment policy just because they own the Word document. Do not let them choose your AI judgment tools just because they own the servers.

Q: "Isn't AI going to get better quickly enough that today's limitations won't matter?"

A: Some limitations will narrow. The contextual judgment gap will likely reduce as models become more sophisticated. The hallucination problem and the ethical reasoning gap are structural. They are features of how probabilistic language models work, not bugs that will be patched. Even the most capable AI systems available today hallucinate with confidence. Plan for hallucination permanently. Build your review gates on the assumption that some percentage of outputs will always be wrong in ways that are not obviously wrong.

Q: "We can't afford to evaluate every vendor this thoroughly."

A: You cannot afford not to, for any tool that touches a protected characteristic, a performance outcome, or a policy decision. The **Vendor Vetting Matrix** takes approximately 90 minutes to complete for a single tool. The average employment discrimination claim costs significantly more than that in management time alone, before legal fees. Apply the full matrix to high-risk tools. Apply a lighter version to low-risk tools. The distinction between high-risk and low-risk is the Zone Sorter from Chapter 1. You already have the framework.

What James Knows Now

James bought a tool he did not understand. That is a fixable problem. What he now has is a framework for making sure it never happens again. And, not just for the next procurement decision, but for every AI tool that enters his people function from any direction.

- He knows that the LLM running the tool is doing math, not thinking.
- He knows that generative AI produces outputs for review and agentic AI executes actions, and that the second category requires a fundamentally different governance response than the first.
- He knows that the hallucination problem is permanent and must be designed around, not hoped away.

And he knows the five questions that separate vendors who understand their tools from vendors who have memorized a reassuring script. Most importantly, he knows where his **Grey**

HR Zone is. Not as an abstract concept, but as a specific list of tools, workflows, and outputs that are currently operating without adequate human oversight. That list is his governance priority. And knowing it is the difference between leading AI and being led by it.

> **Know Your AI** – *Nearly two-thirds of organizations expect to increase AI training offerings — yet most are still in the earliest stages of doing so.*
>
> ~ ATD State of the Industry, 2025

Chapter 3: The Practical AI laybook

Real Implementation for Ethical HR Leaders

CASE IN POINT

Sarah is the Head of People, the Talent Acquisition Lead, the Employee Relations Specialist, and, on some Fridays, the unofficial social secretary for a 180-person construction firm. She is a department of one.

Figure 3.1 - Sarah's Story

Her desk is a physical manifestation of the HR lifecycle: a stack of paper-based onboarding forms, a laptop open to a complex payroll spreadsheet, and a phone that vibrates every ten minutes with a query about holiday allowance or a sub-contractor dispute.

Last year, during a quarterly update, her CEO leaned back and said, "Sarah, I've been reading about how AI is transforming business. We need to get AI into HR by Q3. Let's see some transformation."

Sarah did what any sensible professional would do. She Googled it. The first result was a 40-page whitepaper from a global tech titan detailing how they used a custom-built machine learning cluster and a team of fifty data scientists to reduce attrition in their Silicon Valley headquarters.

Sarah looked at the half-eaten sandwich on her desk and the mounting pile of Right to Work documents.

The gap between the enterprise AI narrative and her reality felt like an ocean.

She closed the laptop.

This chapter was written for Sarah. Not as a consolation prize for not being Google. As a recognition that the gap Sarah felt is the gap most HR professionals feel, and as a practical guide for crossing it, this week, with the budget she actually has.

***

The AI conversation in HR has a visibility problem. The case studies are almost always enterprise. The tools being demoed at conferences are almost always priced for organizations with dedicated HR technology teams and multi-year implementation budgets. The books written about AI in HR are almost always written with a Fortune 500 reader in mind.

The result is that the HR professionals who could benefit most immediately from AI. The ones drowning in repetitive administrative work with no capacity to do the strategic work their organizations actually need, have been effectively written out of the conversation. They see the headlines about AI transforming HR and conclude, reasonably, that it does not apply to them.

It does. In fact, the case for AI in small HR teams is stronger than the case for AI in large ones. And this chapter makes it.

The Small Team Advantage

Here is the thing that enterprise HR leaders rarely admit: being a large organization with a dedicated HR technology function is not actually an advantage when it comes to AI adoption. It is often a significant disadvantage.

Large organizations move slowly. They have procurement committees that take months to approve new tools. They have legacy systems that new AI tools need to integrate with, and integrations that take quarters to build. They have internal politics about which team owns which process, and governance frameworks that were designed for a pre-AI world and have not been updated. They have risk teams that flag everything and legal teams that slow-roll every vendor contract.

Sarah has none of those things. And in the context of AI adoption, that is a competitive advantage.

Why Small HR Teams Can Move Faster

- Sarah can decide to pilot a tool this afternoon. She does not need a business case approved by a steering group. She does not need IT sign-off for a tool she is using on her own device. She can start, assess, and stop in the time it takes a large organization to schedule the first stakeholder alignment meeting. No procurement committee.

- Small HR teams are frequently not locked into enterprise HRIS platforms with complex integration requirements. A solo HR manager using a spreadsheet and a Gmail account can implement an AI-assisted workflow in an afternoon. No legacy system dependencies.

- Sarah does not need to navigate competing priorities between the Talent Acquisition team and the HR Operations team. She is both. Every decision she makes about AI in her function is her decision to make. No internal politics.

- Sarah's CEO is accessible. If she can show him a concrete ROI from a pilot in 30 days, she will have organizational support that a large-company HR leader would spend six months trying to secure through a business case process. Direct line to the decision-maker.

The MVP Mindset

The most important mental shift for small HR teams approaching AI is from "transformation" to "pilot." The enterprise AI conversation is full of transformation language, reimagining, revolutionizing, overhauling. That language is not useful for Sarah. It creates a scale of ambition that makes the first step feel too big to take.

The MVP mindset >>> Minimum Viable Pilot >>> reframes the question entirely. Not "*How do I transform HR with AI?*" but "*What is the single most time-consuming task I do every week that requires no unique human judgment*? And what would happen if AI handled the first draft of it?"

For the scrappy HR leader, AI is not a digital transformation project. It is a better pair of scissors. You are not overhauling the department. You are **removing the friction** that is currently standing between you and the work that actually matters.

Start with one tiny, repetitive friction point. Prove the value and the safety of that one workflow. Bank the time you reclaim as credibility capital which is the evidence that earns you the right to expand to the next workflow. That is how a **Minimum Viable Pilot** becomes a **Humanly Intelligent HR** function.

What 'AI-Ready' Means When You Have No Data Science Team

One of the most common reasons small HR teams delay AI adoption is the belief that they need to get their data in order first. This belief is partially correct and mostly unhelpful.

It is partially correct because AI tools that rely on your historical HR data, such as; workforce planning tools, predictive analytics platforms, sentiment analysis systems, do require reasonably clean, consistent data to function well. If your job titles are inconsistent and your performance ratings have not been reviewed in three years, those tools will not serve you well.

It is mostly unhelpful because the highest-value AI workflows for small teams do not require your historical HR data at all. They require a prompt, a policy document, a job description, or a meeting transcript. The most powerful thing Sarah can do right now does not depend on having a perfect HRIS. It depends on having a clear question and a well-structured prompt.

PRACTICAL DATA HYGIENE FOR SMALL TEAMS

The minimum you need before piloting AI in HR:

For job description generation:
Your current JD template and a list of competencies for the role. That's it.
For policy FAQ chatbot:
Your current employee handbook or policy documents, saved as PDFs or Word documents.
For interview question generation:
Your competency framework, or a description of the role and what success looks like.
For meeting summarization:
A recording or transcript of the meeting.

One practical step before you begin: spend one afternoon cleaning the attic.
Delete the redundant, obsolete, and trivial documents from your HR folders.
If the AI cannot access the 2022 holiday policy, it cannot hallucinate from it.
Clean source documents produce clean AI outputs. That is the whole of data hygiene for small teams.

Figure 3.2 - Practical Data Hygiene

The Highest-Value AI Workflows for Small Teams

The six workflows below were selected on two criteria: they return more time than they cost to set up, and they are all **Zone 1** or **Zone 2** in the **Humanly Intelligence Model**. Meaning they are safe to implement with appropriate human review gates, without requiring the kind of governance infrastructure that only large organizations can build.

Each workflow includes a realistic setup time estimate, a realistic time-saving estimate, its zone classification, and the human review gate required. These are not vendor claims. They are practitioner estimates based on real small-team implementations.

Workflow	Setup Time	Time Saved	Zone	Human Gate Required
Job description generation	2 hrs. setup	2 hrs. saved per role	Zone 1	Light review before posting
Policy FAQ chatbot	3 hrs. setup	5-10 hrs./week	Zone 1-2	Human reviews before any legal query goes live
Interview question generation	1 hr. setup	45 mins saved per role	Zone 1	HR reviews questions before use
L&D content curation	2 hrs. setup	3-4 hrs./week	Zone 1-2	Manager approves the pathway before sharing
Meeting summarizations	30 mins setup	1-2 hrs./week	Zone 1	Minimal review summary before distributing
Onboarding task sequencing	4 hrs. setup	4-6 hrs. per new hire	Zone 2	HR checkpoint before any status-change action

Figure 3.3 - Highest-Value AI Workflows

Workflow 1: Job Description Generation and Bias-Checking

This is the recommended starting point for almost every small HR team. The setup time is minimal, the time saving is immediate and measurable, and the risk of harm is low because the output is a draft that a human reviews before it is posted.

The process: give AI your job title, a list of key responsibilities, and your preferred competency framework (or a description of what success in the role looks like). Ask it to draft a job description. Then ask it separately to check that draft for language that could deter underrepresented candidates. Look for unnecessarily masculine language, degree requirements that are not genuinely necessary, experience requirements that function as proxies for age or background.

The time saving is immediate: from an average of two hours per job description to approximately 20 minutes. For a small HR team hiring 15 to 20 roles per year, that is 25 to 30 hours reclaimed annually from a single workflow change. That is nearly a full working week.

The Grey HR Zone consideration: the final job description must be reviewed by a human before posting. AI bias-checking is a useful starting point, not a definitive audit. The human reviewer is the quality gate, not the AI.

Workflow 2: Policy FAQ Chatbot

For many small HR teams, the single largest drain on time is not complex strategic work. It is the same fifteen questions, asked by different people, every week. What is the annual leave policy? How do I request a flexible working arrangement? What happens if I need to take sick leave beyond three days?

A policy FAQ chatbot, a simple AI tool trained on your employee handbook, can handle these queries without HR involvement. Setup requires uploading your policy documents to an AI tool and configuring it to answer questions **based only on those documents**. This is a Zone 1 workflow for non-legal queries and a Zone 2 workflow for anything that touches entitlements or legal rights.

The critical human review gate: before the chatbot goes live, review its answers to the twenty most common policy questions you receive. Any answer that involves an employee's legal rights, a protected characteristic, or a formal process must be reviewed by you before it reaches an employee. The **Tribunal Test** from Chapter 2 applies here: if the output could be cited in a formal claim, a human must review it first.

Workflow 3: Interview Question Generation

Generating a structured interview question set tailored to a specific role and competency framework typically takes 30 to 45 minutes. AI can produce a first draft in under five minutes. The human reviewer's job is to check that the questions are genuinely competency-based, legally compliant (no questions that could be interpreted as probing protected characteristics), and practically appropriate for the seniority level of the role.

This is a Zone 1 workflow. The AI is drafting. The human is editing and approving. The output is a tool for interviewers, not a decision-making system. The risk is low and the time saving is immediate.

Workflow 4: L&D Content Curation and Learning Pathways

For solo L&D practitioners or HR generalists who own learning alongside everything else, the content discovery problem is real. There is an enormous amount of learning content available, from courses, articles, videos, frameworks, and very little time to curate it intelligently for specific roles or development needs.

AI can dramatically compress the curation process. Give it a role description, a skills gap, and a preferred learning format. Ask it to suggest a 30/60/90-day learning pathway with specific resources at each stage. The output will not be perfect, but it will be a genuinely useful starting point that a human reviewer can refine in 20 minutes rather than building from scratch in two hours.

Workflow 5: Meeting Summarization and Action Extraction

For HR business partners who spend significant time in meetings, such as; performance conversations, team reviews, leadership development sessions, one-to-ones with managers, the administrative work of capturing and distributing notes consumes time that could be spent on the next conversation.

AI transcription and summarization tools can produce accurate, action-oriented meeting summaries in minutes. Ask it to summarize the key points, identify the agreed actions, and flag any mentions of legal risk or policy contradiction. You leave the meeting with the work already 80% finished. The human review gate is simple: read the summary before distributing it, correct any factual errors, remove anything sensitive that should not be in writing, and send.

Workflow 6: Onboarding Task Sequencing – The First Agentic Use Case

This is the most sophisticated workflow in this section, and the one that requires the most careful governance. That is because it is the **first agentic use case most small teams will encounter**. Onboarding task sequencing, the automatic triggering of welcome emails, system access requests, manager meeting scheduling, and first-week checklists, is genuinely valuable for a solo HR manager who is trying to deliver a consistent new hire experience across multiple sites or departments.

The Zone 2 classification is deliberate. Onboarding touches employment status, system access, and the first impression an employee forms of the organization. The human checkpoint requirement is non-negotiable: before any agentic onboarding workflow executes an action that reaches a new employee or changes their system status, a **human must have confirmed the trigger** condition is correct. The Marcus scenario from Chapter 4, the employee who was accidentally offboarded while on sick leave, applies equally to onboarding. If the trigger fires incorrectly, the consequences land on a real person on their first day.

The Scrappy AI Roadmap

The six workflows in the previous section are not a to-do list. They are a menu. The **Scrappy AI Roadmap** is the process for choosing where to start, proving the value, and building from evidence rather than enthusiasm.

Month 1: The Single-Workflow Pilot

Choose one workflow. Not the most exciting one. The one with the highest volume, the lowest risk, and the clearest measurability. For most small HR teams, that is **job description generation** or **policy FAQ** because both produce a concrete, countable output (time saved per job description (JD), number of policy queries deflected) that can be reported to a CEO in a single sentence.

In Month 1, the only goal is to establish a baseline and generate evidence.

- How long does the task currently take?
- How long does it take with AI assistance?
- What is the quality of the output?
- What does the human review gate catch?

Document everything, even informally. You are building the business case for Month 2.

THE MONTH 1 PILOT SETUP
Step 1: Pick your workflow (JD generation recommended for most teams)
Step 2: Set up the tool (most require nothing more than a free account and 30 minutes)
Step 3: Run three to five real tasks through it — actual JDs for actual open roles
Step 4: Time yourself on each one, including review time
Step 5: Compare to your pre-AI time for the same task
Step 6: Note one thing the AI did well and one thing you had to correct
Step 7: At the 30-day mark, write three sentences summarizing what you found
Those three sentences are your CEO update. They are also your permission to proceed to Month 2.

Figure 3.4 - Pilot Setup

Months 2–3: Measuring Impact and Building the Business Case

The internal business case for AI in a small HR team does not need to be a formal document. It needs to be honest, specific, and expressed in language a CEO or CFO responds to. That means numbers, not adjectives.

Not: "AI has improved our recruitment process significantly."

Yes: "AI-assisted JD generation has reduced our time per job description from 90 minutes to 20 minutes. Over the last eight weeks, we have hired for 6 roles. That is 7 hours reclaimed. I am using that time on the skills mapping project we have had on hold since Q1."

In Months 2 and 3, you are doing two things simultaneously: continuing to use and refine the first workflow, and building the evidence base for introducing a second one. By the end of Month 3, you should be able to answer three questions: What did we save? What did we do with the saving? What would we do with a second workflow?

Months 4–6: Expanding to a Second Workflow

The second workflow should be chosen based on evidence from the first, not on enthusiasm. If the JD generation pilot revealed that the biggest remaining time drain is interview question preparation, the second workflow is obvious. If it revealed that policy queries are still consuming two hours a week despite the JD improvement, the FAQ chatbot is the next logical step.

The key discipline at this stage is not to expand faster than your governance capacity. Every new AI workflow requires a human review gate. Every human review gate requires your time. If you are adding workflows faster than you can maintain the review gates, you are not building a Humanly Intelligent HR function. You are building the Grey HR Zone.

The 'Never Automate' List

Every small HR team needs one. Not as a bureaucratic document, but as a personal commitment. A written line of what you will always decide yourself, no matter the pressure, the resourcing, or the vendor promises.

The list is short. The items on it are non-negotiable:

- Any ER case involving a protected characteristic, a mental health disclosure, or a formal grievance.
- Any communication to an employee about their performance, their conduct, or their employment status.
- Any decision that involves a redundancy, a termination, or a change to employment terms.
- Any response to an employee in distress, regardless of the nature of the distress.
- Any situation where the employee has not yet spoken to a human HR professional about the issue.

The resource pressure argument for automating any item on this list is the most dangerous argument you will encounter as a small-team HR professional. Being under-resourced is an argument for more HR headcount. It is not an argument for automating human judgment. If you cannot provide a human response to a Zone 3 situation, the right answer is to escalate, delay, or seek support, not to let the AI handle it.

Making the ROI Case to a CEO Who Has Never Heard of HITL

Your CEO does not need to understand human-in-the-loop AI governance to support your AI investment.

They need to understand three things:

- What it cost
- What it saved, and
- Why you are not automating everything

The first two are easy with Month 1 data. The third requires one clear sentence that frames the governance decision as a business risk management choice, not a technology reluctance:

> *"We are automating the tasks where errors are cheap to fix. We are not automating the decisions where errors cost us legal claims, employee trust, and our reputation as an employer."*

That sentence does not require your CEO to understand the **Humanly Intelligence Model**. It requires them to understand risk and cost...which they already do. Use it.

Cross-Industry Case Study: Professional Services

CROSS-INDUSTRY CASE STUDY

U.S. Professional Services - Solo HR Leader
Organization: 90-person accounting firm
Function: End-to-end HR (recruitment, ER, L&D, compliance, payroll support)

The Problem:

The HR Manager was spending nearly 60% of her week on two types of work:

- Answering repetitive policy questions
- Writing job descriptions for recurring roles

Neither required her expertise.

Both consumed the time she needed to complete a skills framework project the Managing Partner had been requesting for over a year.

She had no dedicated IT support, no additional HR tech budget, and no background in AI implementation.

What She Did

Week 1:
She used a low-cost AI tool to generate job description drafts for the firm's three most frequently hired roles. She reviewed, refined, and saved them as templates.

- Setup time: 4 hours
- Time saved per hire: ~75 minutes

Week 3:
She uploaded the company's employee handbook into an AI tool configured to answer questions based only on that document.

She tested it against the 20 most common policy questions.

- Identified 3 high-risk questions (all related to employee entitlements)
- Routed those directly to herself
- Automated the rest
- Total setup time: ~11 hours across three weeks
- Monthly cost: $50

Outcome

- 14 hours per week reclaimed within the first month
- Skills framework project—delayed for two years—completed within four months
- Zero incorrect policy outputs in the following 12 months

At the three-month mark, she presented the results to the Managing Partner:

"$50 a month, 14 hours a week back and the project is done."

Her AI budget was increased the following quarter.

Total cost: $47 per month for AI subscriptions

Total value: HR transitioned from cost-center administration to strategic partner and not because the HR Manager worked harder, but because AI absorbed the work that was stopping her from doing the work that mattered.

The Lesson:

The ROI was not theoretical. It was 14 hours. The CEO could see it in the work that got done. That is the only business case a small-team HR professional needs.

Figure 3.5 - Cross-Industry Case Study

GREY AREA DECISION FRAMEWORK

Grey Area: The Resource Pressure Trap

This is the grey area that is most specific to small HR teams and the most dangerous one in the book.

For small HR teams, the Grey HR Zone doesn't shrink, it intensifies.

Because the same person is making every decision, the line between automation and judgment is crossed faster, and often invisibly.

There is no colleague to catch it. No second pair of eyes. No governance committee. The resource pressure is real. And the temptation to let it push Zone 3 decisions into Zone 1 is the greatest risk a solo HR professional faces.

The situation: you have five simultaneous grievance cases. You are one person. An AI tool claims it can analyze the evidence and suggest a verdict.

The test: *"Would I be comfortable if this employee found out an AI made, or even heavily influenced, this decision?"*

If the answer is no, or even a hesitant maybe...the workflow stays in Zone 3. Resource pressure does not change the zone classification. It never has.

NEVER automate (regardless of volume or resource pressure):

- ER cases involving protected characteristics
- Mental health disclosures
- Performance management conversations
- Grievance responses
- Any situation where an employee is in distress

The Honest Truth:

Being under-resourced is an argument for more HR headcount. It is never a valid argument for automating human judgment. If you use AI to cut corners on a human life, you are not being scrappy. You are being negligent.

The Grey HR Zone expands every time a small team lets resource pressure push a Zone 3 decision into Zone 1. Your 'never automate' list is the firewall. Keep it short. Keep it non-negotiable.

CHAPTER CHECKLIST

☐ I have identified the single highest-volume, lowest-risk workflow I can pilot with AI this month

☐ I have set a 30-day review point with specific metrics: time saved, quality of output, what the human review gate caught

☐ I have a written 'never automate' list, at least five items, that I have committed to regardless of resource pressure

☐ I have a simple human review gate in place for every AI output that will reach an employee

☐ I have a one-paragraph ROI summary I can share with my CEO or CFO at the 30-day mark

PROMPT LIBRARY & WORKBOOK EXERCISES

PILOT SELECTION PROMPT

[I am a solo HR manager at a [industry] company with [headcount] employees. My three most time-consuming weekly tasks are [list tasks]. For each task, tell me: (1) whether it is a Zone 1, Zone 2, or Zone 3 workflow using the Humanly Intelligence Model, (2) whether AI assistance is appropriate, and (3) if yes, what the human review gate should be.]

ROI PROMPT

[Help me build a one-page business case for using AI in HR for a CEO who is skeptical. My organization is [describe]. The pilot I am proposing is [describe]. Include: time saved per week, annual equivalent, what I will do with the reclaimed time, one comparable example from a similar-sized organization, and the governance safeguards in place.]

WORKBOOK EXERCISE

Time-log your next full working week. Every task, every interruption, every query. At the end of the week, highlight every task that: (a) took more than 30 minutes, AND (b) required no unique human judgment that only you could provide. That highlighted list is your pilot candidate pool. Choose the one with the highest volume. That is your Month 1 workflow.

TEAM EXERCISE (for solo practitioners)

Find a Peer Partner such as another solo HR manager, or a trusted department head in your organization. Share your pilot plan with them before you start. Ask them one question: *"If you were an employee, would you feel dehumanized by this AI use case?"* Use their answer to tune your human review gate. If their answer is yes, or even uncertain, your governance is not yet ready. Go back to the **Zone Sorter**.

SKEPTIC'S CORNER FAQ

Q: "I don't have time to implement AI on top of everything else I'm doing."

A: This is the ultimate paradox. You do not have time because you are not yet using the tools that create time. The workflows in Section 3.2 were chosen specifically because they return more time than they cost to set up. Job description generation takes two hours to set up and saves two hours per role from day one. If you hire 15 roles this year, that is 25 hours returned for a two-hour investment. The time you do not have is the reason to start, not the reason to wait.

Q: "What if I set something up and it goes wrong?"

A: Every workflow in this chapter has a human review gate built in. Nothing in Section 3.2 reaches an employee without you seeing it first. The risk is not a bad decision that harms someone. It is a bad draft that you catch before it goes anywhere. That is a recoverable situation. The unreviewable agentic system with no human checkpoint is the risk to manage. These workflows are not that.

Q: "My CEO wants AI everywhere immediately."

A: Chapter 8 gives you the full change management language for managing upward on this. The short version: show your CEO the Humanly Intelligence Zone Sorter. Point to your never-automate list. Then say: "Moving too fast into Zone 3 is a legal and reputational liability. I am moving as fast as the governance allows, and faster than that is not a competitive advantage. It is a risk we do not need to take." That is a sentence a CEO who has been to a conference will understand.

What Sarah Does on Monday Morning

Sarah closes this chapter with something she did not have when she Googled "AI for HR" and got back a case study from Google or Gartner. She has a starting point that is sized for her reality.

She knows that the first step is not a platform procurement, a vendor demo, or a board presentation. It is a time log. One week. Every task. Then a single question: which of these tasks takes more than 30 minutes and requires no unique human judgment that only she can provide?

She knows that job description generation is almost certainly the answer. She knows it will take two hours to set up. She knows it will save her two hours per role. And she knows that she will not send a single AI-generated output to a hiring manager without reading it first. And, this is because the human review gate is not optional, and it is not a burden. It is the thing that makes the whole system trustworthy.

The enterprise AI conversation will continue without her. The conferences will keep showcasing the tools that require seven-figure implementation budgets and dedicated HR technology teams. And Sarah will keep doing what small-team HR professionals have always done: making it work with what she has, better than anyone thought possible. The only difference now is that she has AI helping her do it.

> **The Scrappy HR Playbook** – *67% of HR professionals say their organizations were not proactive in upskilling employees to work with AI. 51% say enhanced training is their organization's top need. Only 1 in 4 HR professionals played a leading role in AI implementation.*
>
> ~ SHRM, From Adoption to Empowerment, 2025

Chapter 4: Agentic AI

When AI Stops Drafting and Starts Doing

CASE IN POINT

Marcus is the CHRO of a mid-size financial services firm. He has been in HR for eighteen years. He has navigated mergers, restructures, a global pandemic, and two hostile employment tribunal cases. He is, by any measure, an experienced and capable people leader.

Figure 4.1 - Marcus' Story

Last quarter, a Workday configuration team embedded within HR operations implemented what they called the "Seamless Transition Engine," an automated offboarding workflow. Marcus wasn't consulted.

On paper, it was a masterpiece of efficiency. The moment a resignation was logged in the HRIS, the system triggered a cascade of actions: revoking building and network access, initiating payroll offboarding, generating COBRA notices, and sending exit survey links to the employee's personal email. All of it. Automatically. Without a human approving each step.

It worked perfectly for dozens of departures, until it didn't.

Last month, a data entry error had entered a departure date in the wrong employee's record. She was on extended sick leave, recovering from surgery. The Seamless Transition Engine did not see a human being in a hospital bed. It saw a data trigger.

Her laptop was locked within the hour. Her email credentials were revoked while she was in a telehealth appointment, which she could no longer access. The exit survey arrived in her personal inbox while she was recovering. Her payroll was flagged for offboarding processing.

She called Marcus, crying.

The error took eleven days to fully reverse. Not because the systems were difficult to update, but because the organization had no documented process for reversing an automated offboarding that should never have been initiated. Nobody had designed for the failure mode because nobody in HR had known the workflow existed.

Marcus is a good CHRO. He is not an outlier. He is a preview of what happens when agentic AI enters organizations faster than the governance frameworks designed to contain it.

This chapter starts there.

***

Every chapter in this book matters. This one is different.

The previous three chapters have prepared you to diagnose what is broken, understand the technology, and implement AI wisely in your people function. This chapter names the shift that changes the risk profile of everything. The move from AI that drafts content for humans to review, to AI that executes actions that affect people's working lives.

Agentic AI is not a future technology. It is already running in most organizations. It is running in payroll systems, HRIS platforms, onboarding workflows, and compliance tracking tools. It was often implemented by IT or operations, without HR's input, without HR's knowledge, and without anyone stopping to ask: what happens when this gets something wrong?

This chapter gives you the framework to answer that question, before the answer becomes a phone call from an employee who is crying.

What Agentic AI Is and Why It Changes Everything

The distinction between generative AI and agentic AI was introduced in Chapter 2. This section builds on that foundation with the specific implications for HR governance.

Generative AI produces an output and stops. You prompt it, it generates, and you decide what to do with the result. Human judgment closes every loop. Even if the human is under-resourced, even if the review is cursory, there is a moment where a human hand touches the output before it reaches another person. However brief, that moment matters.

Agentic AI does not stop at the output. It continues through to the action. It is given a goal >>> complete this process, manage this workflow, handle this task, and it takes the steps required to achieve that goal autonomously. It does not wait for human approval at each step. It does not pause to ask if the conditions have changed. It executes.

THE CRITICAL DISTINCTION

GENERATIVE AI: Prompt → Output → Human reviews → Human acts

The AI produces. The human decides.

AGENTIC AI: Trigger → AI executes a sequence of actions → Outcome reaches a person

The AI decides. The AI acts. The human finds out.

In HR, the difference between these two models is the difference between a bad draft you catch and a wrong action that has already affected someone's employment, pay, or access.

Figure 4.2 - Critical Distinction

Current Agentic AI Applications in HR

Agentic HR workflows are not hypothetical. They are operational in most organizations right now. The most common include:

- Triggered by departure date entry. Revokes system access, initiates payroll changes, schedules equipment collection, sends exit surveys. Often implemented by IT as a "process efficiency" initiative. Offboarding automation.
- Triggered by hire confirmation. Creates system accounts, sends welcome emails, schedules induction meetings, assigns compliance training. Often runs before background checks are confirmed. Onboarding task sequencing.
- Triggered by enrollment window opening or closing. Sends reminders, processes elections, escalates non-responses. Often operates without HR visibility into individual employee status. Benefits enrollment management.
- Triggered by training deadlines. Sends reminders, escalates to managers, flags non-completions for HR review, in some cases triggers formal non-compliance processes automatically. Compliance training tracking.
- Triggered by calendar events or performance cycle milestones. Initiates pay review workflows, routes approvals, in some cases processes pay changes without final human confirmation. Payroll and pay review scheduling.

Why Agentic AI Is Already in Your Organization

The majority of HR leaders who are reading this chapter did not implement the agentic workflows currently running in their people function. They were implemented by IT, by payroll providers, by HRIS vendors, by operations teams. They were described as "automation," "process improvements," or "system upgrades." HR was either not consulted or was consulted at a level that did not surface the governance implications.

This is shadow automation, the workflows running in the background of your people function that nobody in HR formally approved, nobody in HR is currently monitoring, and nobody in

HR will know about until something goes wrong. Most HR leaders don't realize these workflows are running until someone calls them crying.

This is not a failure of HR leadership. It is a structural problem with how AI implementation has been governed in most organizations: technology teams make implementation decisions, and HR inherits the consequences. The accountability gap between technical ownership and people impact is where the Marcus scenario lives.

The first thing this chapter is asking you to do is not to design new governance. It is to find out what is already running. You cannot govern what you cannot see. And right now, most HR leaders cannot see the full landscape of agentic workflows operating in their people function.

The Amplification Problem

When a human makes an error in an HR process, the error is typically contained. One person receives incorrect information. One process produces a wrong outcome. The damage is real, but it is limited by the pace at which humans work.

When an agentic system makes an error, it executes that error at the speed and scale of the system. Every employee in the trigger condition receives the wrong action simultaneously. Every record that matches the flawed logic is processed incorrectly before anyone notices. By the time a human identifies the problem, the error may have already reached dozens or hundreds of people.

In plain terms, the action has already happened by the time HR is aware of it.

This is the amplification problem. It is not a reason to avoid agentic AI. It is a reason to design every agentic HR workflow on the assumption that it will eventually produce a wrong output AND to have a remediation process designed before that happens, not after.

And what you are designing for is only going to grow.

The agentic workflows running in your organization today are early-stage. What comes next is a different order of magnitude entirely.

> Analyst *Josh Bersin's HR 2030 research* envisions a future where agentic systems handle the full recruitment lifecycle from sourcing, screening, assessment, interviewing, offers, and onboarding. All this while simultaneously delivering personalized learning, managing real-time performance data, and flagging pay equity issues without waiting to be asked. HR Business Partners, in that vision, transition from process administrators to agent managers, steering the systems, not running the transactions.

Bersin estimates that HR teams in this future will be 30 to 40 percent smaller. That is not a threat. It is a design challenge. And it is the challenge that makes the governance work in this chapter not a compliance exercise but a strategic imperative. Because the organizations that arrive at HR 2030 with no human oversight architecture will not be more efficient. They will be ungovernable.

That is what this chapter is built to prevent.

The Agentic HR Workflow Audit

Before you can govern agentic AI in your people function, you need to know what you are governing. The Agentic HR Workflow Audit is a structured process for mapping every automated workflow currently operating in your HR ecosystem. This includes the shadow automation running without HR's knowledge or approval.

This is not a future scenario. In many organizations, shadow automation is already affecting employee records, pay, access, and compliance...right now. The audit is how you find it before it finds you.

This audit requires no technical knowledge. It requires conversations, documentation review, and the willingness to ask questions that some IT and operations colleagues may find uncomfortable.

How to Identify Every Agentic Workflow

Start with three conversations and one document review:

1. Ask specifically: "What automated workflows in our HRIS, payroll system, or any connected platform execute actions that affect employee records, access, pay, or communications, without requiring a human to approve each action?" The word 'automated' is important. You are not asking about reports or dashboards. You are asking about workflows that do things. Your IT team or HRIS administrator.

2. Ask the same question. Payroll systems are frequently the source of agentic workflows that HR leadership is unaware of, particularly around pay changes, offboarding triggers, and benefits processing. Your payroll provider.

3. Ask what 'out of the box' automated workflows are enabled in your current configuration. Vendors often activate default automations during implementation that are not explicitly surfaced to HR during onboarding. Your HRIS vendor.

4. If accessible, review the change logs for your HRIS and payroll system for the last 12 months. Look for any workflow or automation that was activated or modified. Cross-reference against HR's awareness of those changes. Your system change logs.

The Five Questions for Each Agentic Workflow

This is where governance either holds or breaks. For every agentic workflow you identify, ask these five questions. Document the answers. Any question you cannot answer is a governance gap. And a governance gap in an agentic system is not a theoretical risk. It is a very active exposure.

1. If nobody in HR authorized it, that is your first finding. If it

has not been reviewed in the last 12 months, that is your second. Who authorized this workflow and when was it last reviewed?

2. What event causes it to fire? What happens if the trigger data is incorrect? Is there a mechanism for the workflow to recognise that the trigger condition is wrong? What triggers the workflow, and what are the failure conditions?

3. At what point does a human review what the workflow is about to do before it does it? If the answer is "there isn't one," this workflow is operating in the Grey HR Zone without governance. Where is the human checkpoint, and is there one?

4. Not if it goes wrong. When. Does a documented remediation process exist? Who owns it? How quickly can the system be reversed? What is the remediation process when the workflow produces a wrong outcome?

5. If yes, the governance requirements are non-negotiable. Any agentic workflow that touches these areas without a human checkpoint is operating outside the boundaries of the Humanly Intelligence Model. Does this workflow touch a protected characteristic, a pay decision, or an employment status?

The Agentic Workflow Registry

The output of your audit is the **Agentic Workflow Registry**. This is a living document that maps every agentic workflow in your people function against the five audit questions. It is not a one-time exercise. It is a governance tool that is **reviewed and updated every quarter**, and updated immediately whenever a new agentic workflow is implemented or an existing one is modified.

The Registry serves three purposes:

- It gives HR visibility into what is running

- It provides the audit trail that regulators and employment tribunals increasingly expect to see, and
- It is the governance foundation that makes everything else in this chapter possible.

The template below is your starting point. ***Appendix C*** *contains the full downloadable version.*

Designing Human-in-the-Loop (HITL) for Agentic Systems

Knowing what is running is the first step. The second is designing the human judgment back into the workflows that are currently operating without it. This section gives you three models for doing that — and the decision matrix for choosing which model to apply to which workflow.

The Three HITL Models

HITL MODEL 1: REVIEW BEFORE EXECUTION
What it is: A human reviews and approves the workflow action before it executes.
When to use it: Zone 2 workflows where the action directly affects an individual's employment status, system access, or pay.
Example: Offboarding access revocation — HR confirms the resignation is active before the workflow revokes access.
The trade-off: Slower. Requires HR availability. But the only model appropriate for high-stakes, irreversible actions.

Figure 4.3 - HITL Model 1

HITL MODEL 2: EXCEPTION-BASED REVIEW
What it is: The workflow executes automatically in normal conditions but pauses and escalates to a human when it encounters an edge case.
When to use it: Zone 1–2 workflows where the majority of cases are straightforward, but specific conditions (employee on leave, open ER case, recent grievance) require human judgment.
Example: Compliance training escalation, the workflow sends standard reminders automatically, but pauses and escalates to HR if the employee has an open ER case or has recently flagged a well-being concern.
The trade-off: Requires the workflow to be designed with explicit edge case recognition. More technically complex to build, but manageable for most HRIS platforms.

Figure 4.4 - HITL Model 2

HITL MODEL 3: POST-EXECUTION AUDIT
What it is: The workflow executes automatically, but every action is logged and reviewed by a human on a regular cadence.
When to use it: Zone 1 workflows where the actions are low-stakes, reversible, and the volume makes pre-execution review impractical.
Example: Benefits enrollment reminders >>> the workflow sends reminders automatically; HR reviews the log weekly to catch any anomalies.
The trade-off: Does not prevent errors from reaching employees. Catches them after the fact. Appropriate only for workflows where errors are genuinely low-harm and easily reversible.

Figure 4.5 - HITL Model 3

The HITL Decision Matrix

Use this matrix to assign the appropriate HITL model to every workflow in your **Agentic Workflow Registry**.

Workflow	Zone	HITL Model	Human Checkpoint Requirement
Onboarding task sequencing	**Zone 2**	Exception-based	If the background check is not cleared, or the employee's start date changes, pause and escalate
Benefits enrollment reminders	**Zone 1**	Post-execution audit	Log all actions; review weekly for errors or missed enrollments
Compliance training tracking	**Zone 1–2**	Exception-based	If the completion deadline passes with no action, escalate to manager and HR before penalty triggers
Offboarding access revocation	**Zone 2**	Review before execution	HR must confirm resignation/termination is active before any access change executes
Pay review scheduling	**Zone 1**	Post-execution audit	Log scheduling actions; human reviews before any pay change is communicated
Contract renewal notification	**Zone 1**	Post-execution audit	Log notifications; HR reviews before any contract action is triggered
Termination or redundancy	Zone 3	**NEVER agentic**	Human decision-making and communication are required at every stage. No exceptions.
Mental health / ER escalation	Zone 3	**NEVER agentic**	Human must initiate, manage, and close. AI may support documentation only.

Figure 4.6 - HITL Decision Matrix

How to Retrofit HITL into Existing Workflows

Most of the agentic workflows already running in your people function were not built with HITL in mind. Retrofitting human judgment into an existing workflow is almost always possible and rarely as technically complex as it sounds.

The most common retrofit approach is the '**pause gate**': inserting a step in the workflow that sends a notification to a named HR person, requires confirmation before the next step executes, and times out with an escalation if confirmation is not received within a defined window. This can typically be configured in your HRIS without custom development.

The conversation with IT is the most important non-technical action you can take after completing the Workflow Audit. HR does not need to understand the code. HR needs to articulate the governance requirement clearly:

"This workflow affects employment status. Before it executes any action that changes someone's access, pay, or employment record, I need a notification that allows me to confirm the action is correct. If I do not confirm within 24 hours, the workflow should pause and escalate to me, not proceed."

That sentence does not require technical knowledge. It requires an understanding of what the workflow does and what the governance requirement is. Both of those things are now in your **Agentic Workflow Registry**.

The Escalation Trigger – Designing AI That Knows When to Stop

The most sophisticated element of HITL design for agentic systems is the escalation trigger: the condition under which the workflow recognizes that it has encountered something outside its design parameters and stops to ask a human.

The basic escalation triggers every agentic HR workflow should include:

- Sick leave, maternity/paternity leave, compassionate leave, or any leave type associated with a protected characteristic. If the trigger condition fires while an employee is on protected leave, the workflow must pause. Employee on protected leave.

- If an employee has an open employee relations case, a formal grievance, or a recent formal complaint, any automated action affecting their employment requires human review before execution. Open ER case or grievance.

- If the data triggering the workflow is inconsistent with other data in the system; for example, a departure date that conflicts with an active sick leave record, the workflow must pause and flag the anomaly for human review. Data anomaly.

- If the workflow is about to execute an action it has not performed before for this employee, or an action that falls outside the normal distribution of previous executions, it should flag for review. First execution of a new pattern.

The Agentic Opportunity – What HR Should Be Asking IT to Build

This chapter has, rightly, focused on the governance requirements of agentic AI. But governance without opportunity is just restriction. This section names the agentic workflows that represent genuine, high-value, low-risk opportunities for HR functions of every size, And, gives you the language to brief your IT or AI team to build them safely.

Four High-Value Agentic Workflows Worth Prioritizing

When a hire is confirmed and the background check clears, an onboarding system can take over the administrative flow..

- It creates system access.
- Orders equipment.
- Schedules induction.
- Assigns compliance training.

All in the right sequence.

With human checkpoints at every step that affects employment status. HR focuses on the welcome and cultural integration. The system handles the logistics. The time savings are significant. The governance requirement is straightforward. New hire onboarding task sequencing.

- Benefits enrollment windows are time-bound, repetitive,

and administratively intensive. An agentic workflow that sends tiered reminders, tracks completion, and escalates non-responses to HR saves significant administrative time with low governance complexity. The post-execution audit HITL model is appropriate for most organizations. Benefits enrollment reminders and deadline management.

- Mandatory training completion is a legal requirement in most regulated industries. An agentic workflow that tracks completion, sends reminders at defined intervals, escalates to managers when deadlines approach, and flags persistent non-completions to HR is high-value, low-risk, and appropriate for exception-based HITL. The exception: any escalation that triggers a formal process must involve human review before the formal step is taken. Compliance training tracking and escalation.
- The offboarding process involves multiple coordinated actions across HR, IT, payroll, and facilities. An offboarding automation engine can coordinate those actions, But only after a human has confirmed the departure is active, voluntary, and not subject to any live ER proceedings. The **Review Before Execution model** is mandatory for any offboarding workflow. The Marcus scenario is the consequence of skipping it. Exit process coordination.

How to Write an HR Brief for an IT or AI Team

The most common reason HR's governance requirements are not built into agentic systems is not technical. It is communication. HR leaders do not speak the language IT teams need to build to specification, and IT teams do not ask the governance questions HR needs them to ask.

The template below gives you a plain-English brief structure that any IT or AI team can work from:

THE HR AGENTIC WORKFLOW BRIEF TEMPLATE

Workflow name: [What is this workflow called?]

Purpose: [What problem is this solving? What outcome does it produce?]

Trigger: [What event causes this workflow to start?]

Actions: [List every action the workflow takes, in sequence]

Zone classification: [Zone 1 / Zone 2 / Zone 3]

HITL model required: [Review Before Execution / Exception-Based / Post-Execution Audit]

Human checkpoint: [At what step must a human confirm before the next action executes?]

Escalation triggers: [Under what conditions must the workflow pause and alert HR?]

Remediation process: [If the workflow executes an error, what is the process to reverse it?]

Named HR owner: [Who is accountable for this workflow's outcomes?]

Review cadence: [When will this workflow be reviewed, and who will review it?]

Sample Completed Form

THE GREY HR ZONE SERIES · CHAPTER 4

HR Agentic Workflow Brief

Complete one form per registered workflow. Retain in the Agentic Workflow Registry.

WORKFLOW IDENTITY

Workflow name	Automated Offer Letter Generation and Delivery
Purpose	Eliminates manual drafting time for standard offer letters by generating compliant, role-specific offer documents and routing them for approval upon candidate selection. Produces a finalized, sent offer within 24 hours of verbal acceptance.
Trigger	Recruiter marks candidate status as "Verbal Offer Accepted" in the ATS.

ACTIONS

1. Pull candidate record, role data, and compensation details from ATS
2. Select the correct offer letter template based on employment type, location, and FLSA classification
3. Populate all variable fields and run a compliance check against current state and federal requirements
4. Route draft offer letter to hiring manager inbox for review
5. Upon hiring manager approval, route to HR for final sign-off
6. Deliver countersigned offer letter to candidate via secure e-signature platform
7. Log completion and timestamp all actions in the Agentic Workflow Registry

GOVERNANCE CLASSIFICATION

Zone classification	Zone 2
HITL model required	Review Before Execution
Human checkpoint	Step 4. The hiring n ↓ ,er must approve the populated draft before the document advances to HR or is sent to the candidate. No

Figure 4.7a - HR Agentic Workflow

RISK AND REMEDIATION

Escalation triggers	• Compensation fields fall outside the approved band for the role level • Candidate location triggers a jurisdiction not covered by existing templates • Compliance check flags a state-specific requirement with no current template match • Hiring manager does not respond within 48 hours of routing
Remediation process	If an offer letter is sent in error, HR Owner initiates a recall through the e-signature platform, sends a direct written retraction to the candidate within two business hours, documents the error in the Registry with root cause notation, and submits a workflow modification request within five business days.

OWNERSHIP AND REVIEW

Named HR owner	Director of Talent Acquisition
Review cadence	Quarterly. Reviewed by the Director of Talent Acquisition and the HR Compliance Lead. Annual review includes Legal sign-off on template currency and jurisdictional compliance.

AWB-2024-004 · v1.0 Entered by: T. Williams · Date: 03/14/2024

Figure 4.7b - HR Agentic Workflow

This brief is not a technical specification. It is a governance specification. It tells the IT team what HR needs the system to do from a human oversight perspective. It is the bridge between HR's accountability for people outcomes and IT's accountability for system design.

The most important line in the brief is the last one: the named HR owner. Every agentic workflow that affects your people must have a human owner in HR who is accountable for its outcomes. Not a team. Not a function. A person. That is what accountability means in the age of agentic AI.

Cross-Industry Case Study: U.S. Healthcare Network

CROSS-INDUSTRY CASE STUDY

Organization: A 2,500-person US healthcare network, VP of People Operations
Function: HR Operations and Onboarding

The Problem:

The organization had implemented an AI-powered onboarding system eighteen months prior. The system was designed to trigger the moment a new hire's start date was confirmed in the HRIS: it automatically issued offer documentation, sent system access requests to IT, enrolled the new hire in mandatory compliance training, and scheduled induction sessions.

The problem emerged gradually. The compliance team began noticing that new hires were accessing the system and completing compliance training before their background checks had fully cleared. In three cases, employment documentation had been issued to candidates whose background checks subsequently revealed disqualifying information. In each case, the organization had to reverse the employment offer, a significantly more complex and legally sensitive process than declining before documentation was issued.

When the VP of People Operations investigated, she discovered that the onboarding workflow had no background check gate. The system had been implemented to trigger on start date confirmation, not background check confirmation. Nobody in HR had been involved in the trigger design decision.

What They Did:

The VP convened a cross-functional review with IT, the HRIS vendor, and the compliance team. The review mapped every trigger in the onboarding workflow, fourteen in total, of which HR had been aware of eight.

For each trigger, the team asked the five Agentic Workflow Audit questions. The result was a redesigned workflow with a mandatory 'background check confirmed' gate inserted before any employment documentation was issued or system access granted. The gate required explicit HR confirmation, not just a system status change, before the workflow proceeded.

The team also discovered three additional agentic workflows that had been implemented by the HRIS vendor as default configurations and had never been reviewed: an automated probation completion notification that went to employees before managers had confirmed the outcome, a contract renewal reminder that triggered without checking for active ER proceedings, and a compliance escalation that sent formal non-compliance notices to employees without HR review.

All three were modified to include human review gates before the chapter's HITL framework was formally implemented.

Outcome:

Zero compliance violations in the 18 months following the redesign

Onboarding completion rate improved by 23% as a by-product of the audit. The mapping process revealed several workflow steps that were firing in the wrong sequence and causing new hires to receive conflicting instructions

Figure 4.8a - Cross-Industry Case Study

The Agentic Workflow Registry became the organization's standard governance tool for all subsequent HR system implementations

Key Insight:

The audit revealed 14 agentic workflows, of which HR had been aware of 8. The 6 unknown workflows were all implemented by IT or the HRIS vendor as default configurations. They were not secret. They simply never surfaced to HR because nobody had asked the right questions.

The right question is: "*What does this system do autonomously, and what happens when it gets something wrong?*"

Figure 4.8b - Cross-Industry Case Study

GREY AREA DECISION FRAMEWORK

Grey Area: Employment Status Changes

This is the most consequential grey area in the agentic AI landscape, and the one that produces the Marcus scenario.

The test for any agentic workflow touching employment status:

"Does this workflow change something an employee would need to be told about by a human?"

If yes, human checkpoint required before execution. No exceptions.

NEVER execute autonomously:

- Terminations or redundancies – the communication must come from a human voice
- Contract changes or amendments – every change requires human confirmation
- Pay changes – any increase, decrease, or correction
- Employment status changes – full-time to part-time, permanent to fixed-term, active to leave
- Sensitive leave approvals – compassionate leave, mental health leave, any protected leave type

CAN execute with human review gate (Review Before Execution model):

- Probation completion notifications – after manager confirms outcome
- Contract renewal reminders – after HR confirms no live ER proceedings
- Standard pay review scheduling – not pay changes, only scheduling
- Training completion certificates – after completion is confirmed

The escalation rule – non-negotiable:

Any agentic workflow that encounters an edge case **must pause** and **escalate** to a named HR person.

Edge cases include: employee on any form of leave, open ER case, recent formal grievance, data anomaly in the trigger condition.

The Grey HR Zone in agentic AI is not the action the workflow takes when everything goes right. It is the action the workflow takes when everything goes wrong and nobody designed for that.

CHAPTER CHECKLIST

☐ I have identified every agentic workflow currently running in my people function, including those implemented by IT or vendors without HR input

☐ I have completed the Agentic Workflow Registry for at least three workflows, using the five audit questions

☐ I have identified which workflows have no human checkpoints; these are my immediate governance priorities

☐ I have a written escalation protocol for at least one agentic workflow, naming the HR owner and the escalation path

☐ I have a conversation scheduled with my IT or operations team to discuss HITL requirements for existing and future agentic workflows

PROMPT LIBRARY & WORKBOOK EXERCISES

WORKFLOW AUDIT PROMPT

[I want to audit the agentic AI workflows in my HR function. Generate a set of discovery questions I can use in interviews with my IT team]

HITL DESIGN PROMPT

[I am going to describe an agentic HR workflow. Help me design a human-in-the-loop protocol for it using the three HITL models: Review Before Execution]

WORKBOOK EXERCISE

Draw a simple flowchart of one agentic workflow currently running in your organization. Label every step the workflow takes automatically. Mark with a ✓ every point where a human currently reviews the action before it executes. Circle every point where an error could reach an employee without human review. Each circle is a governance gap. The circle closest to the highest-stakes action is your immediate priority.

TEAM EXERCISE: THE FAILURE MODE

Take one agentic workflow and run a failure mode exercise with your HR and IT teams. Ask two questions:

1. 'What is the worst realistic thing this workflow could do if the trigger data is wrong?'
2. 'If that happened right now, what would we do, and how long would it take to reverse?'

The gap between those two answers is your remediation design brief.

SKEPTIC'S CORNER FAQ

Q: "Agentic AI sounds dangerous, should we just avoid it?"

A: You cannot avoid what is already inside the building. Agentic logic is the foundation of modern HRIS platforms, payroll systems, and IT infrastructure. If you try to ban it, your teams will simply use it without your oversight. The choice is not between using agentic AI and not using it. It is between governed automation and shadow automation. Avoiding conversation does not prevent the exposure. It guarantees that you find out about it the way Marcus did.

Q: "This is really an IT problem, not an HR problem."

A: IT is responsible for the code. HR is responsible for the consequences. If an automated system incorrectly offboards an employee on sick leave, the IT team does not receive the phone call. You do. If an agentic workflow violates a labor law or discriminates against a protected group, the "IT made me do it" defense will not hold in court or in the court of public opinion. If it touches a human life, it is an HR problem. The workflow touches your people. The compliance risk falls on HR. That accountability requires HR to be in the room when these systems are designed.

Q: "We don't have the technical knowledge to audit these systems."

A: You do not need to read code to audit a workflow. You only need to understand three things: the trigger, the actions, and the human checkpoint, or lack of one. If you can explain your organization's disciplinary policy, you have enough analytical capability to audit an agentic workflow. The five discovery questions in Section 4.2 require zero technical knowledge. They require professional judgment. That is exactly what HR has.

What Marcus Knows Now

Marcus could not have prevented the Marcus scenario with better intentions. He needed better visibility. He needed to know that the workflow existed. He needed to have asked, at some point, the question that no one had thought to ask: "What does our system do automatically, without HR's involvement, that affects the employment status of our people?"

He now has a framework to answer that question: the Workflow Audit, the Registry, the five questions, the three HITL models, and the Agentic Workflow Brief template. None of it is technically complex. All of it is governmentally essential.

Agentic AI will not stop expanding in HR. The workflows will become more sophisticated. The actions will become more consequential. The organizations that navigate this well will not be the ones that adopted agentic AI most enthusiastically. They will be the ones that were most deliberate about where the humans stayed. That deliberateness is the Grey HR Zone, applied to the most powerful AI that has ever entered your people function.

Agentic AI – 82% of HR leaders plan to use some form of agentic AI within their functions by May 2026. By 2030, 50% of current HR activities will be AI-automated or performed by AI agents.

~ Gartner, Prepare for the Future of AI Agents in HR, December 2025

Chapter 5: Ethical AI in HR

The How, Not Just the Why

CASE IN POINT

Priya is a Talent Acquisition Director at a technology company. She has been in talent acquisition for twelve years. She cares deeply about fairness. She has led inclusion initiatives. She has championed diverse hiring panels. She thought she was building an equitable recruitment process.

Figure 5.1 - Priya's Story

Eight months ago, her organization implemented an AI screening tool. The vendor's pitch was compelling: faster screening, consistent criteria, and the phrase that appeared in the contract, a "bias-free guarantee." Priya read that guarantee carefully. She asked about the methodology. The vendor provided a reassuring two-page summary. She signed.

Last week, a data analyst in another department ran a routine demographic report on recruitment outcomes. She shared it with Priya almost as an afterthought.

The pattern was unmistakable. The AI screening tool was filtering out women over 40 at three times the rate of any other demographic group. Not slightly more. Three times. Eight months of active hiring. Dozens of roles. Hundreds of candidates were affected.

Priya sat with the data for a long time before she picked up the phone.

The vendor's "bias-free guarantee" did not mention the protected characteristics it had been tested against. It did not mention the demographic composition of the training data. It did not mention what happens when the guarantee turns out to be wrong.

This chapter explains what Priya does next. And, more importantly, how to make sure this never happens again.

***

The AI ethics conversation in HR has a problem. It has been dominated, for too long, by the **why**.

HR leaders know why algorithmic bias matters. They have read the headlines about the Amazon recruiting tool that penalized CVs mentioning women's colleges. They have seen the research on facial recognition tools that misidentify darker-skinned faces at higher rates. They have sat through the conference sessions, read the white papers, and absorbed the message: AI can perpetuate and amplify human bias at machine speed and scale.

What they have not been given is the **how**.

- How do you find bias in an AI tool you are already using?
- How do you fix it without a data science degree?
- How do you prevent it from reappearing?
- How do you build the kind of ethical AI governance that your organization can actually maintain and not as a

one-off audit, but as an operating standard?

This chapter answers those questions. It gives you a **seven-step bias audit process** that any HR leader can run, a plain-English explanation of how to interpret the results, a framework for responding when you find what Priya found, and the blueprint for an Ethical AI Charter that will hold your people function accountable for getting this right.

> Ethical AI is where the **Grey HR Zone** becomes most visible and most consequential. This is where judgment cannot be outsourced, and where the cost of getting it wrong is not operational. It is human.

If you are not auditing for bias, you are not managing risk. You are hoping it doesn't surface. This chapter is about making sure it does, on your terms, before someone else finds it for you.

Understanding Algorithmic Bias – Plain English

Algorithmic bias is not a malfunction. It is a consequence. It is what happens when an AI system produces outcomes that systematically disadvantage certain groups. And, not because the system was designed to discriminate, but because the data it was trained on, or the variables it was given, reflected historical inequities that the system then learned to replicate.

Understanding how bias enters an AI system is the foundation of knowing how to find and address it. There are three primary mechanisms in HR AI:

Historical Data Bias

Most AI tools used in HR, particularly in recruitment and performance management, are trained on historical data. That data reflects historical decisions. And historical hiring and promotion decisions, in most organizations, were made by humans who operated within systems that were not equitable. A tool trained to identify "successful" candidates based on the characteristics of historically successful hires will learn to replicate whatever patterns, conscious or unconscious, shaped those hires.

This is the Garbage In, Garbage Out problem applied to people decisions. In practice, it means your AI is learning from a version of your organization that may never have been fair to begin with. A recruitment AI trained on a technology company's historical hiring data might learn to deprioritize candidates who took career breaks, not because career breaks indicate poor performance, but because the candidates who were historically hired had fewer career breaks, often because they were less likely to have primary caregiving responsibilities. The bias is not in the code. It is in the data. And the data reflects a world that was not fair.

Proxy Discrimination

Proxy discrimination is arguably the most insidious form of algorithmic bias because it operates through variables that appear neutral. A proxy variable is an input that correlates with a protected characteristic without being that characteristic directly.

Postcode is a classic proxy variable. It correlates with race and socioeconomic status in most urban contexts. So does the name of the university attended, the gap between degree completion and first employment. And, as Priya's case illustrates, age-related patterns in work history. A screening tool that uses any of these variables as positive or negative signals is engaging in proxy discrimination, even if the tool was designed to be "blind" to the protected characteristics themselves.

The critical insight about proxy variables: they are often invisible until you look for them deliberately. The Scandinavian retail case study in this chapter illustrates exactly this. The tool appeared to be scoring on writing quality. It was actually disadvantaging non-native speakers. The proxy was hidden in a seemingly neutral input.

Feedback Loop Bias

Feedback loop bias occurs when an AI system's outputs become its own future training data. In HR, this most commonly appears in performance management and talent identification tools. If an AI tool consistently rates employees in a particular demographic group lower, those lower ratings become part of the performance record. That performance record then informs future AI-assisted decisions about development, promotion, and retention. The bias compounds with every cycle.

Feedback loop bias is the slowest-moving and hardest-to-detect form of algorithmic bias in HR. It does not appear in a single demographic report. It accumulates over time, often invisibly, until the disparity becomes large enough to be unmistakable, at which point significant harm has already been done.

Why 'Bias-Free' Is a Marketing Claim, Not a Technical Specification

No AI tool used in HR is bias-free. This is not a controversial statement. It is a technical fact. Every tool trained on human-generated data inherits some degree of the patterns in that data. Every tool that uses proxy variables, even ones designed to reduce bias, carries the risk of proxy discrimination. Every tool that feeds its own outputs back into its training data carries the risk of feedback loop bias.

"Bias-free" is a marketing claim. It tells you about a vendor's confidence in their product or their willingness to make assurances they cannot keep. It does not tell you what protected characteristics were tested, what demographic compositions were represented in the training data, what the acceptable threshold for disparity was, or when the last audit was conducted. Any vendor who claims their tool is bias-free has told you they will not be useful partners in addressing the bias that exists.

The HR Leader's Bias Audit – A Step-by-Step Process

The seven steps below give you a complete bias audit process for any AI-assisted HR workflow. You do not need a data science degree to complete them. You need to be a curator of fairness. Someone who looks at the data your tools produce and asks, consistently and without flinching, whether the outcomes are equitable.

You need access to your HRIS or vendor data, a spreadsheet, and the willingness to ask questions that may produce *uncomfortable answers*. The audit is designed for HR leaders, not analysts. The language is plain. The tools are accessible. The output is actionable.

STEP 1: Define What Fairness Means for This Specific Workflow

Before you look at a single number, define what a fair outcome looks like for this workflow.

Ask: *Should the selection rate for this workflow be equal across demographic groups?*
Or: Is there a legitimate reason why selection rates might differ? (e.g., a physical fitness requirement for a role with genuine physical demands)

Write your fairness definition down before you look at the data.
This matters because if you define fairness after you see the numbers, you will unconsciously define it in a way that explains away disparities.
Your pre-defined fairness standard is your audit benchmark.

Figure 5.2a - Bias Audit (Step 1)

STEP 2: Request Demographic Disaggregation Data

Ask your vendor or HRIS for the selection rates of this workflow, broken down by every protected characteristic available.

The data you need: for each demographic group - how many applied (or were assessed), how many were selected (or passed), and what the selection rate was.

If your vendor cannot provide this data, that is itself a finding. A vendor who cannot disaggregate outcomes by demographic group cannot meaningfully assure you their tool is fair.

If your HRIS does not capture demographic data, that is a data collection gap to address. You cannot audit for bias against characteristics you are not tracking.

Figure 5.2b - Bias Audit (Step 2)

STEP 3: Run the Four-Fifths Rule

The four-fifths rule (also called the 80% rule) is the most widely used non-technical method for detecting potential discriminatory impact. It was developed by the US Equal Employment Opportunity Commission and is recognized in UK and EU employment law contexts as a useful screening tool.

How it works:

1. *Identify the demographic group with the highest selection rate. This is your comparison group.*
2. *Calculate 80% (four-fifths) of that selection rate.*
3. *If any other demographic group has a selection rate below that threshold, the disparity warrants investigation.*

What it tells you: whether a disparity exists that is large enough to be legally and practically significant.
What it does not tell you: whether the disparity is caused by bias, or what the cause is. That requires Steps 4 and 5.

Figure 5.2c - Bias Audit (Step 3)

The worked example below shows the four-fifths rule applied to a recruitment screening tool, using the demographic data from Priya's situation:

Group	Applied	Selected	Selection Rate	4/5ths of Highest	Flag?
Men under 40	120	60	**50%**	- (this is the highest)	**No**
Women under 40	80	36	45%	40% (4/5 of 50%)	**No**
Men over 40	60	22	37%	40% (4/5 of 50%)	Borderline
Women over 40	40	7	**17.5%**	40% (4/5 of 50%)	**YES - investigate**

Figure 5.3 - Recruitment Screening Tool

In this example, women over 40 have a selection rate of 17.5% against the highest group's 50%. The four-fifths threshold is 40%. The disparity is not borderline. It is significant, and it requires immediate investigation.

STEP 4: Look for Proxy Variables

A proxy variable is any input to the AI tool that correlates with a protected characteristic.

How to identify them: for each input the tool uses, ask - "Could this input be systematically different for any protected group, for reasons unrelated to job performance?"

Common proxy variables in HR AI:

Postcode or location - correlates with race, socioeconomic status
University attended - correlates with socioeconomic status, potentially race
Employment gaps - correlates with gender (caregiving), disability, and age
Years of experience - correlates with age
Writing complexity or style - correlates with first language (native speaker advantage)
Commute distance - correlates with socioeconomic status, disability

What to do when you find one: assess whether it is a genuine job-relevant requirement or a proxy. If it is a proxy, it should be removed from the model inputs.

Figure 5.2d - Bias Audit (Step 4)

STEP 5: Test with Synthetic Candidates — The Secret Shopper Method

Synthetic candidate testing is a manual bias check that any HR leader can run without technical skills.

How it works:

1. Create two or more candidate profiles that are identical in all job-relevant characteristics (qualifications, experience, skills).
2. Vary only the characteristics associated with a protected group (e.g., name suggesting a different ethnicity, graduation year suggesting age, employment history suggesting a career break).
3. Submit both profiles to the AI tool.
4. Compare the outputs.

If the tool produces materially different outputs for profiles that are identical in job-relevant characteristics, the difference is attributable to the varied characteristics. That is evidence of bias.

This test is most useful for recruitment screening tools. It is less applicable to tools that rely primarily on your organization's internal data.

Figure 5.2e - Bias Audit (Step 5)

STEP 6: Document Findings and Create a Remediation Plan

Every finding from your audit must be documented, regardless of severity.

Your documentation should include:

The workflow was audited and the date of the audit
The demographic data reviewed and the methodology used
The disparities identified, with specific numbers
The proxy variables identified
The synthetic candidate test results, if conducted
The remediation actions proposed and the responsible owner for each
The timeline for remediation
The date of the next scheduled audit

This documentation is your defense if your organization is ever challenged on the fairness of an AI-assisted decision. It demonstrates that you took the obligation seriously, that you identified issues, and that you acted on them.

Figure 5.2f- Bias Audit (Step 6)

STEP 7: Set a Monitoring Cadence

A bias audit is not a one-off exercise. It is the beginning of a monitoring program.

Minimum cadence: quarterly for high-stakes workflows (recruitment, performance management, pay decisions). Annually for lower-stakes workflows.

Trigger events that require an immediate unscheduled audit:

Any change to the AI tool's model or training data
Any change to the demographic composition of your candidate or employee population
Any formal complaint or legal challenge relating to fairness
Any significant change in outcome rates that was not anticipated

The monitoring cadence is what separates a one-time compliance exercise from a genuine ethical AI governance program.

Figure 5.2g - Bias Audit (Step 7)

When You Find Bias – What to Do

Priya has the data. She knows what the tool has been doing. The question now is not whether to act. It is how to act, in what order, and with what urgency.

Finding bias is not a failure. Ignoring bias is a failure. When your audit reveals a disparity, you must act with the same urgency Marcus should have shown in his offboarding crisis. The response has five components. They are sequential. Do not skip steps or reverse the order.

Step 1: Pause the Workflow and Protect Affected Individuals

The first action is immediate and non-negotiable: pause the AI-assisted workflow. Not "review it" or "monitor it closely." Pause it. Until you understand what the tool has been doing and have a remediation plan in place, every output it produces is suspect.

For recruitment tools, this means reverting to manual screening for all active roles while the audit is completed. For performance tools, it means flagging all recent AI-assisted ratings for human review before any decision based on those ratings is actioned.

Simultaneously, identify the individuals most likely to have been materially affected. In Priya's case, this means identifying women over 40 who were screened out during the eight-month period the tool was active and who might have been shortlisted under a fair screening process. These individuals require a specific remediation response, not just an internal process fix.

The Internal Escalation Path

Bias findings of this magnitude require immediate escalation through a specific sequence. The order matters because each step requires the previous one:

1. HR leadership, your CHRO or equivalent. They need to know immediately. This is a legal and reputational risk, not a process improvement matter.

2. Legal counsel, before you communicate anything externally, internally, beyond a small group, or to the vendor. Legal needs to advise on your obligations and exposure before you generate any documentation that could be discoverable.

3. The CEO or relevant executive, after legal has advised. This is a board-level risk in a significant organization. The CEO should not find out from a journalist or a regulator.

4. Your Data Protection Officer or Privacy team, if the bias finding involves the processing of special category data

or has GDPR implications, the DPO needs to be involved before any external communication.

5. The vendor, after you have legal advice and know what you are asking for.

Vendor Accountability

The vendor has a "bias-free guarantee" in the contract. That guarantee has demonstrably not been met. The conversation with the vendor is not a customer service call. It is a formal contractual discussion, and it should be conducted with legal counsel involved.

What to demand from the vendor:

- A full technical disclosure of the bias testing methodology that underpinned the 'bias-free guarantee', including the protected characteristics tested, the training data demographic composition, and the acceptable disparity thresholds used.
- An explanation of why women over 40 were not identified as an at-risk group in the pre-deployment testing.
- A remediation plan with a specific timeline for modifying the model.
- Clarity on liability - what the vendor's contractual responsibility is for the discriminatory outputs their tool produced.

What remediation looks like in practice: at a minimum, the model inputs need to be reviewed for proxy variables, the training data needs to be audited for demographic representation, and the tool needs to be retested against the affected demographic groups before it is reactivated. Any vendor who cannot commit to this is not a responsible partner.

Legal Obligations

Whether Priya's organization has a legal obligation to notify anyone, as in the affected candidates, a regulator, or an employment authority, depends on jurisdiction, the severity of the impact, and whether affected individuals have been materially harmed. This is not a question this book can answer definitively, because the answer varies significantly by location and circumstance.

What this book can tell you: do not make this assessment yourself. This is precisely the situation in which **legal counsel earns their value**. The obligation to notify, the timing of notification, and the content of any notification all carry legal implications that must be managed by qualified professionals.

Rebuilding Trust with Affected Groups

The most difficult part of the bias response is not the legal process or the vendor conversation. It is the communication to the people who were harmed.

Women over 40 were screened out of Priya's hiring process at three times the rate of any other demographic group, for eight months, by a tool that was running without adequate oversight. Some of them applied for roles they were qualified for and did not hear back. Some of them may have concluded that they were not competitive for that type of role. The harm is real, even if no single individual can be proven to have been the decisive victim.

The communication framework for affected groups must be honest, specific, and offer a concrete remedy. Not "we have identified an issue with our recruitment process and are working to address it." That is a statement designed to minimize legal exposure, not to restore trust.

Where affected candidates can be identified and meaningfully contacted, the communication should name what happened, acknowledge the impact, and offer a specific opportunity to be reconsidered for relevant roles by a human reviewer. That is not a guarantee of employment. It is a genuine attempt to restore the fairness that was taken away.

Building an Ethical AI Charter for Your People Function

A bias audit tells you what is happening now. An **Ethical AI Charter** tells you what you are committed to ensuring happens going forward. It is the governance foundation that makes bias auditing a systematic practice rather than a crisis response.

Every HR team that uses AI in any people process needs one. Not because a regulator requires it, though regulatory requirements in this space are moving fast, but because the absence of a charter is the absence of a standard. **And without a standard, there is no accountability.**

What an Ethical AI Charter Is

An Ethical AI Charter is a one-to-two page document that articulates your HR function's commitments around the use of AI in people decisions. It is not a legal document. It is not a policy document in the traditional sense. It is a public commitment, to your employees, your candidates, and your leadership, about the principles that govern how AI is used in your function and how you will hold yourselves accountable to those principles.

The distinction between a charter and a policy is important.

- A policy tells people what they must do.
- A charter tells them what you stand for.

Both are necessary. The charter comes first, because it defines the values that the policies must serve.

The Five Commitments Every Charter Should Include

The table below gives you the five core commitments of an Ethical AI Charter, what each commitment means in practice, and the evidence you should maintain to demonstrate that you are living by each one.

Commitment	What We Will Do	Evidence We Will Maintain
Bias Prevention	We will audit every AI tool that influences a people decision for demographic disparity before deployment and annually thereafter.	Bias audit log; vendor audit report; four-fifths rule results
Human Oversight	No AI system will make a final employment, performance, or pay decision without human review. AI advises. Humans decide.	HITL protocol for each AI tool; named reviewer for each workflow
Employee Transparency	We will inform employees when AI is used in processes that affect them and provide a clear path to request human review.	Employee-facing AI disclosure; review request process
Data Privacy	We will not use employee data for AI model training without explicit consent and will conduct a DPIA for every new AI tool.	DPIA register; consent framework; data minimization policy
Continuous Monitoring	We will review AI tool performance for bias and accuracy on a quarterly cadence and update this charter annually.	Quarterly review calendar; charter review date; named charter owner

Figure 5.4 - Ethical AI Charter Five Elements

How to Write One in a Half-Day Workshop

An Ethical AI Charter does not require a consultant, a six-week project, or an executive retreat. It requires a half-day with the right people in the room and a clear facilitation structure.

Who should be in the room: your CHRO or HR Director, your HR operations lead, your legal or compliance contact, your privacy or data protection lead, and, if possible. one or two employee representatives who can bring the "*how does this land for the people we serve*" perspective.

The facilitation structure:

1. **Opening (30 minutes):** Share two or three case studies of algorithmic bias in HR, including the one from this chapter. The goal is not to create fear but to create shared understanding of what is at stake.

2. **Current state mapping (45 minutes):** List every AI tool currently used in your people function. For each one, rate your current governance maturity against the five charter commitments. Where are you strong? Where are you exposed?

3. **Commitment drafting (60 minutes):** Using the five commitments as a framework, draft specific, measurable language for each one that reflects your organization's

context. What does 'bias prevention' mean specifically for your tools, your roles, your population?

4. **Evidence and accountability (30 minutes):** For each commitment, agree on the evidence you will maintain and the person accountable for maintaining it. A commitment without a named owner is not a commitment. It is a wish.

5. **Review and sign-off (15 minutes):** Read the draft charter aloud as a group. Does it say what you mean? Would your employees recognize it as a genuine commitment or a corporate statement? Revise until the answer is yes.

Getting Executive Sign-Off

The **Ethical AI Charter** needs executive endorsement to be meaningful. Not because HR cannot operate without it, but because the commitments in the charter have implications for procurement, for system design, for vendor management, and for the legal and compliance functions. And, those implications require executive authority to action.

The business case language that consistently works with executive teams is not the ethical argument. It is the risk and competitive argument:

THE EXECUTIVE BUSINESS CASE FOR AN ETHICAL AI CHARTER
"The EU AI Act, emerging US state legislation, and UK Equality Act case law are all moving toward mandatory proactive auditing of AI systems used in employment decisions. Our Ethical AI Charter positions us ahead of that requirement rather than behind it.
The average employment discrimination claim costs significantly more than the investment required to build this governance framework. We are not spending money to do the right thing. We are spending money to avoid a significantly higher cost.
Our Ethical AI Charter is also a talent acquisition asset. Candidates increasingly ask about AI use in hiring. Employees increasingly ask about AI use in performance management. A public commitment to ethical AI governance differentiates us as an employer that takes this seriously.
We are asking for your endorsement of a two-page document and a quarterly 90-minute review meeting. In return, we are managing a legal, reputational, and talent risk that is growing every quarter."

Figure 5.5 - Executive Business Case Language

Cross Industry Case Study: Global Retail Organization

CROSS-INDUSTRY CASE STUDY

Organization: A Global retail organization, 12,000 employees, Head of People Analytics

Function: Performance management

The Problem:

The organization had implemented an AI-assisted performance scoring tool across its regional network. The tool aggregated multiple inputs to produce a quarterly performance score for each employee: manager ratings, customer satisfaction data, sales metrics where applicable, and written self-assessment.

Over three performance cycles, the Head of People Analytics noticed a persistent pattern: employees in one regional office were consistently scoring lower than equivalent performers in other regions, despite similar sales metrics and manager ratings. The office in question had a significantly higher proportion of non-native English speakers than the rest of the network.

What They Did:

The team ran a retrospective four-fifths rule audit on the performance scores, disaggregated by first language (native vs. non-native English speaker). The disparity was clear and significant. Non-native speakers were scoring an average of 18 percentage points lower than equivalent native speakers.

The team then conducted a proxy variable investigation on the tool's inputs. The written self-assessment component, heavily weighted because it correlated with high scores in historical training data, was producing systematically lower scores for employees who wrote in a less complex or less fluent English style.

The tool was not measuring performance. It was measuring written English fluency and rewarding it as if it were a performance indicator.

The assessment was discriminatory on the grounds of national origin, a protected characteristic in the jurisdictions where the organization operated.

Outcome:

- The written self-assessment component was removed from the AI scoring model pending a full redesign
- The model was rebuilt using only the objective, non-language-dependent inputs: sales metrics, customer satisfaction scores, and structured manager ratings
- All performance scores from the three affected cycles were retrospectively reviewed by a human panel
- Two employees who had been passed over for promotion based on their AI-generated performance scores were retrospectively reviewed and subsequently promoted

The organization implemented the **Ethical AI Charter framework** from this chapter within six months of the finding.

The Lesson:

Proxy variables are often invisible until you look for them deliberately. The written self-assessment appeared to be a performance measure. It was functioning as a language fluency measure.

The four-fifths rule surfaced the disparity. The proxy variable investigation identified the cause. Neither step required a data science degree. Both required the willingness to ask uncomfortable questions about data that appeared neutral.

Figure 5.6 - Cross Industry Case Study

GREY AREA DECISION FRAMEWORK

Grey Area: Bias vs. Legitimate Differentiation

Not every demographic disparity in AI output is illegal bias. But all of it requires examination.

This is one of the most important distinctions in ethical AI governance, and one of the most misused.

The test: Is the differentiating factor directly and probably related to the requirements of the role?

Example of legitimate differentiation:

- A tool that screens out candidates without a valid driving license for a role that requires driving.
- The requirement is genuine, measurable, and directly related to the role. The disparity it produces, if any, is legally defensible.

Example of proxy discrimination:

- A tool that scores candidates lower for 'culture fit' without a defined, measurable culture fit framework.
- 'Culture fit' is not a job requirement. It is a preference. And preferences, when encoded into AI systems, reliably reproduce the demographics of whoever defined the culture.

The rule: When in doubt, apply human review before any disparity-affected decision is actioned.

If you cannot explain to a rejected candidate exactly what the AI assessed and why it is directly relevant to the role, the assessment is not defensible.

The documentation rule:

- Document every disparity you find and what you did about it. This is your defense if challenged.

The organization that found bias, investigated it, and remediated it is in a significantly better legal position than the organization that found nothing because it never looked.

> The **Grey HR Zone** in bias: it is not the disparity that defines the zone. It is the absence of a process for examining it.

CHAPTER CHECKLIST

☐ I have requested demographic disaggregation data for at least one AI-assisted HR workflow and reviewed it against the four-fifths rule.

☐ I have run or scheduled a four-fifths rule check on our AI recruitment tool, using the worked example in this chapter as a guide..

☐ I have identified the proxy variables most likely to create bias in our specific organizational context.

☐ I have a written escalation process for when bias is detected >>> naming the sequence, the people involved, and the actions at each step.

☐ I have drafted, commissioned, or scheduled the development of an **Ethical AI Charter** for our HR function.

PROMPT LIBRARY & WORKBOOK EXERCISES

BIAS AUDIT PROMPT

[I want to conduct a bias audit on our AI [recruitment / performance / L&D] tool. Walk me through the four-fifths rule step by step, using this example data: [paste your demographic data]. Identify any demographic groups whose selection rate falls below the four-fifths threshold. For each flagged group, suggest the most likely proxy variables that could explain the disparity, and recommend the next investigation steps]

CHARTER PROMPT

[Help me draft a one-page Ethical AI Charter for our HR function. It should include our commitments on: bias prevention, human oversight, employee transparency, data privacy, and continuous monitoring. Tone: practical and non-legalistic. Each commitment should include a specific, measurable statement of what we will do and what evidence we will maintain. Audience: our employees, candidates, and executive leadership.]

WORKBOOK EXERCISE

Name one AI tool currently used in your people function. Write down everything you know about how it was tested for bias before purchase. Include: what protected characteristics were tested, who conducted the testing, what the acceptable disparity threshold was, and when the last audit was conducted. Be honest about the gaps. That gap list is your audit starting point. Each gap is also a question for your next vendor conversation.

TEAM EXERCISE: THE PROTECTED CHARACTERISTIC WALK-THROUGH

For one AI-assisted workflow in your people function, ask your team to trace the journey of four specific hypothetical candidates through the tool:

1. A 22-year-old male recent graduate applying for a management trainee role

2. A 45-year-old woman returning from a five-year career break for caregiving

3. A 30-year-old candidate with a non-western name and equivalent qualifications

4. A 55-year-old internal applicant with 20 years of tenure applying for a promotion

At each stage of the workflow, **ask**: *where do their paths diverge*? *What input is driving that divergence*? The divergence points are your proxy variable candidates. The inputs driving them are your **investigation priorities**.

SKEPTIC'S CORNER FAQ

Q: "Our legal team says we don't have to audit AI tools proactively."

A: Your legal team is advising you on current minimum requirements. That is their job. Your job is to consider where those requirements are heading and what the cost of being behind the curve will be. Regulatory requirements in the EU AI Act, emerging US state legislation, and UK Equality Act case law are all moving toward mandatory proactive auditing. The organization that has been auditing voluntarily for two years when mandatory requirements arrive is in a fundamentally different position than the one that was waiting for the mandate. Getting ahead of this is not just ethical. It is strategically sensible.

Q: "If we find bias, are we legally obligated to report it?"

A: This depends on jurisdiction, severity, and whether affected individuals have been materially harmed. The answer is genuinely not simple, which is exactly why you need legal counsel involved in your bias response protocol before you find anything. One thing is certain: avoiding the audit because you are afraid of the answer is not a defence. It is an admission of negligence. The organization that found nothing because it never looked is in a significantly worse legal and ethical position than the one that found bias, documented it, and acted on it.

Q: "Our vendor says they audit for bias continuously."

A: Ask them to show you four things: the methodology, the frequency, the protected characteristics tested, and the last audit report. All four. Not a summary. Not a policy document. The actual results. If they cannot provide all four, their assurance is a marketing claim, not a technical guarantee. And, you have just learned something important about what kind of partner they will be when you find something they were supposed to have caught.

What Priya Does Now

Priya has the audit process. She has the escalation path. She has the vendor accountability framework. She has a communication approach for affected candidates. And she has the **blueprint for the Ethical AI Charter** that will ensure this never happens again on her watch.

She also has something harder to name but more important than any of the frameworks: the knowledge that a "bias-free guarantee" is not a guarantee. That ethics in AI is not a vendor responsibility. That the HR leader who deploys a tool and trusts the marketing language is the HR leader who will eventually be sitting with data like hers, trying to work out what to do next.

Bias is not the failure. Not looking for it is.

The organizations that get ethical AI right in HR are not the ones that found the perfect tool. They are the ones **who built the governance** to examine every tool, question every guarantee, audit every output, and act on what they find. That is what **Humanly Intelligent HR** looks like in **the age of algorithmic systems**.

> **Ethical AI / Bias** – *89% of CEOs believe CHROs should have a central role in driving long-term organizational growth. Yet HR remains largely absent from the AI governance decisions shaping that future.*
>
> ~ Society for Industrial and Organizational Psychology (SIOP), 2025

Chapter 6: Recruitment eimagined

Restoring Human Judgment to a Polluted Pipeline

CASE IN POINT

Tom is a Talent Acquisition Manager at a professional services firm. He has been recruiting for eleven years. He is good at it. He knows what good looks like.

Figure 6.1 - Tom's Story

Last month he posted a role. Within 72 hours he had 340 applications.

His AI screener shortlisted 12. He reviewed all 12. He interviewed 6. He hired nobody.

Not one of the 6 candidates he interviewed could demonstrate the core capability the role required. The shortlist was technically compliant. Every candidate matched the keyword criteria the screener was configured to prioritize. But keyword matching and job readiness are not the same thing. They never were. And, an AI screener cannot tell the difference.

Tom went back to the original 340. He screened them manually for over three weeks. He found four candidates who were genuinely strong. Two of them had been screened out in the first pass because their applications were written in plain English rather than job description language. One had described leading a complex change program without using the words 'change management.' The other had described building client relationships without using the word 'stakeholder.' Both were excellent. Neither made the AI's shortlist.

Tom is not sure whether the AI saved him any time overall. He is sure that if he had not gone back to the pile, he would have failed to fill a role his business needed filled.

This is not an argument against AI in recruitment. It is an argument for knowing exactly what AI can and cannot do. And, designing your process around that knowledge rather than around the vendor's promise.

***

Recruitment is the most consequential thing an organization does. Every hire shapes the team, the culture, the capability, and the trajectory of the business. The decision to bring a person into an organization, to offer them a livelihood, to make a commitment to them, is not an administrative process. It is a human one.

And yet recruitment, more than any other HR function, has been handed to AI with the least governance, the least scrutiny, and the least understanding of what the technology is actually doing. The efficiency promise was compelling. The volume problem was real. The shortcuts were easy to take.

But here is the thing that most of the efficiency conversation misses: recruitment has not become more efficient. It has become more automated. And those are not the same thing.

Analyst *Josh Bersin* has documented what HR leaders have been feeling for years without the data to name it. The talent shortages driving organizations toward automated screening are not cyclical. They are structural. In what Bersin calls the post-industrial age, workforce scarcity is the permanent condition and the organizations winning the talent game are the ones that have stopped screening for credentials and started hiring for skills. That pressure is real. The shortcuts it produced are understandable. And the damage they have done is significant.

The result is what Chapter 1 named as **recruitment pollution**: a feedback loop in which AI generates applications and AI screens applications, with the human on both sides increasingly alienated from the process that is supposed to connect them. The best candidates opt out. The weakest candidates, the ones most skilled at gaming keyword systems, make it through. Quality collapses. And talent acquisition leaders like Tom spend three weeks undoing what the algorithm did in 72 hours.

The problem is not AI in recruitment. The problem is the belief that recruitment is a filtering problem instead of a judgment problem. When you frame it as filtering, AI looks like the obvious solution. When you frame it correctly, as one of the most **consequential series of human judgments** your organization makes, the question changes entirely. The question is not how to automate more of it. It is how to ensure that human judgment is present at every stage where it matters.

> Recruitment is not a narrow **Grey HR Zone**. It is one of the largest Grey Zones in the entire people function, because almost every step involves judgment disguised as process.

The zone sorter from Chapter 1 applied to recruitment reveals an uncomfortable truth: the vast majority of stages where AI is currently operating belong in Zone 2 or Zone 3. AI has been handed Zone 3 decisions and told they are Zone 1 tasks. That is where the quality collapse comes from.

This chapter is about breaking that loop. Not by removing AI from recruitment, but by being deliberate about where it belongs and where it does not. Navigating it well is the difference between building a great team and building a compliant shortlist.

The Recruitment Pollution Problem

To understand what needs to be fixed, you need to understand what broke it. The recruitment pollution problem did not arrive suddenly. It accumulated gradually, as AI tools were introduced on both sides of the application process without anyone designing for the interaction between them.

The AI Application Flood

AI application generation tools have fundamentally changed the economics of applying for work. Applying for a job used to cost something: time, effort, the cognitive labor of articulating why you are the right person for this specific role. That cost was a natural filter. Candidates applied for roles they were genuinely interested in, or at least genuinely suited for, because applying for a role they were clearly wrong for was not worth the effort.

That filter is gone. AI application generators allow a candidate to apply for a role in minutes; a tailored CV, a bespoke cover letter, and language optimized for the specific job description. The cost of applying is now effectively zero. The result is that volume has exploded while signal quality has collapsed. A hiring manager receiving 340 applications is not receiving 340 serious expressions of interest. They are receiving 340 AI-optimized documents, many from candidates who have never considered whether they are genuinely suited for the role.

This is not a moral failing on the part of candidates. It is a rational response to a system that was already optimized against them. When applicants discovered that AI screeners were filtering on keywords, they responded by keyword-optimizing their applications. The arms race, the feedback loop created, was inevitable once AI entered on both sides of the process.

> When both sides use AI, the process stops being human selection and becomes system matching. We did not remove bias. We removed friction, and with it, judgment.

The AI Screening Paradox

AI screening tools are optimized for pattern matching. They identify candidates whose applications resemble the characteristics of previously successful candidates, or whose language matches the language of the job description. This is a useful capability for filtering obvious mismatches at high volume. It is a dangerous capability for identifying genuine potential.

The candidates who most resemble previous successful hires are not necessarily the candidates who will perform best. They are the candidates who have been trained in the same environments, speak the same professional language, and present in the same way. A tool optimized to find "more like our best" will reliably reproduce the demographic and background profile of the existing team. That is not talent identification. It is pattern cloning.

The candidates who write about their experience in their own words, who use human language rather than job description language, who describe what they did rather than what they were 'responsible for', are the candidates most likely to be filtered out. They are also frequently the candidates who think most independently. AI screening optimizes for algorithmic fluency, not human capability. And those are not the same thing.

The Candidate Trust Crisis

The best candidates . . . the ones with genuine options, the ones who do not need to apply for every role that appears in their feed, are opting out of bot-led recruitment processes. This is not speculation. It is visible in application withdrawal rates, offer acceptance declines, and the growing body of candidate sentiment research that identifies AI-heavy processes as a significant negative signal about an organization's culture.

When a candidate applies for a role and receives no human communication for three weeks, then receives an automated rejection with no specific feedback, they conclude that the organization that issued that rejection. They share that conclusion with their networks. They post it in the professional communities where your next strong candidate is deciding whether to apply. The reputational cost of a dehumanized recruitment process is not visible on a time-to-hire dashboard. It is visible in the quality and diversity of your pipeline six months later.

The **Grey HR Zone** in candidate trust is the space between "we use AI to manage volume" and "candidates feel processed rather than considered." That space is where employer brands are built or destroyed.

The Quality Collapse

Tom's situation is not unusual. Talent acquisition leaders across sectors are describing the same pattern: AI screening tools that produce technically compliant shortlists that do not contain the talent the role requires. The tools are working exactly as designed. The design was wrong.

A shortlisting tool optimized for keyword matching will produce a shortlist of skilled keyword matchers. A shortlisting tool trained on historical hiring data will produce a shortlist of candidates who resemble historical hires. Neither of these outcomes is the same as a shortlist of the candidates most likely to excel in the role.

The gap between what AI screening measures and what successful performance requires is the quality collapse. The tool is not broken. The assumption that screening is the same as selecting is broken. And in most organizations, that assumption is still running the process.

The Hybrid Recruitment Model

The solution to recruitment pollution is not to remove AI from the process. It is to be specific and deliberate about what AI is actually good at in recruitment, and to design human judgment back into every stage where capability, potential, and fit are being assessed.

The **Hybrid Recruitment Model** applies the **Humanly Intelligence Framework** directly to talent acquisition. It asks, for every stage of the recruitment process: is this a task where AI can own the output, a task where AI and human judgment should work in tandem, or a decision that must belong entirely to a human?

Recruitment Stage	Zone	Human Requirement
Application acknowledgement	Zone 1 — Automate	Instant automated confirmation with personalized touchpoint option for shortlisted candidates
Initial eligibility check	Zone 1 — Automate	Verifying minimum stated requirements (qualifications, right to work). Not capability or potential.
Keyword and competency signal detection	Zone 2 — Collaborate	AI surfaces patterns; human reviewer assesses whether signals reflect genuine capability or keyword gaming
Diversity pipeline monitoring	Zone 2 — Collaborate	AI tracks demographic composition; human reviews for proxy variable risk and takes corrective action.
Interview scheduling	Zone 1 — Automate	Fully automated. Low risk, high timesaving, candidate-friendly
Shortlisting decisions	Zone 3 — Human	Human judgment only. AI may surface, but a named human must decide who advances. Always.
Competency-based interview	Zone 3 — Human	Structured human conversation. AI may assist with question generation, not scoring or assessment
Culture fit assessment	Zone 3 — Human	Undefined culture fit scored by AI is proxy discrimination; human judgment with a validated framework only.
Final hiring decision	Zone 3 — Human	Human. Always. Non-negotiable. This is the employment relationship.

Figure 6.2 - Hybrid Recruitment Model

What AI Should Own in Recruitment

The administrative infrastructure of recruitment is genuinely AI-appropriate. Acknowledging applications instantly. Sending scheduling links. Confirming interview times. Maintaining candidate communication cadences. Tracking pipeline status and generating reporting on diversity metrics and time-to-stage conversion. None of these tasks require human judgment. All of them consume human time that could be spent on the stages that do require judgment.

The discipline required is not adding AI to these tasks. Most organizations have already done that. The discipline is ensuring that AI does not creep beyond them. Every time someone proposes extending AI's role in the recruitment process.

The question is not *"Would this be efficient?"* It is, *"Is this a task, or is this a judgment?"*

Tasks can be automated. Judgments cannot. And the most expensive mistake in AI-assisted recruitment is consistently the same one: treating a judgment as if it were a task.

Designing the Human Moment Back In

The most valuable design intervention in a polluted recruitment process is not a new tool. It is a **new principle**: every candidate who could plausibly be right for this role should encounter a human being before they are eliminated.

This does not mean screening 340 applications manually. It means designing the process so that human judgment enters before the final shortlisting decision, not after it.

AI surfaces. Human decides.

That sequence must be non-negotiable, and it must be embedded in the process design rather than left to the discretion of an overstretched recruiter.

THE CANDIDATE EXPERIENCE AUDIT

Before redesigning your process, map what candidates actually experience.

For one recently completed recruitment campaign, trace the following:

- *From application submission to first human contact: how many days?*
- *How many candidates were eliminated without any human ever viewing their application?*
- *At what point in the process does a candidate first encounter a human being?*
- *What is the quality of the communication a rejected candidate receives?*
- *If a strong candidate applied in plain English (without keyword optimization), would your process surface them?*

The answers to these questions tell you where your human moments have been removed. They are your **process reform** priorities.

Figure 6.3 - Candidate Experience Audit

The AI-Proof Job Description

The job description is the first conversation you have with every candidate who will ever apply for your role. It is also, in the world of AI application generators, a document that is actively being reverse-engineered by AI tools to produce applications that match it.

If your job description is built on keywords and credential requirements, you are training the application generators to optimize for keywords and credentials. And your screener will select for the same. The JD is training your screener to screen out good candidates. Most hiring teams do not realize this is happening because they never trace the logic from the JD to the application to the shortlist. This section does it for you.

The AI-proof job description is not written to defeat AI generators. It is written to attract:

- Human thinking

- To communicate what the role is genuinely about

- What capability it requires, and
- What a person who would be good at it would actually look like

When you do that, you make the role harder to game and more likely to attract candidates who are genuinely suited to it.

Why Most Job Descriptions Are Training AI to Screen Out Good Candidates

The typical job description is a list. A list of responsibilities, followed by a list of requirements, followed by a list of "nice to haves." It is written by extracting elements from the previous version of the JD, cross-referencing against a job family framework, and adding whatever the hiring manager requested in the briefing call. The result is a document that describes a category of person rather than a specific role.

When an AI application generator encounters this document, it extracts the keywords and phrases and builds an application around them. When an AI screener encounters the applications it produces, it selects the ones that most closely mirror the original keyword list. The feedback loop is complete. The JD produced a set of keywords. The applications reflected those keywords. The screener selected for those keywords.

Nobody in this loop assessed whether any of those keywords were genuinely related to what makes someone excellent in this role.

The question to ask of every element in your job description: "*If a candidate matched this requirement perfectly but was terrible at the job, would I know from their application*? And *if a candidate was outstanding at the job but expressed it differently, would my screener find them?*"

If the answer to the second question is no, the JD is training your screener to screen out good candidates.

The Competency-First JD Framework

The Competency-First framework restructures the job description around what the role requires a person to be able to do, rather than what qualifications or experience they should have accumulated. It is more honest, more specific, and significantly harder to keyword-game than a traditional credential-based JD.

Section	What to Include	What to Avoid
Role purpose	What this person is trying to achieve in 12 months. One or two sentences.	*A list of duties that reads like a job advert from 2005.*
Core competencies	3–5 competencies with behavioral indicators. What does good look like?	*20+ bullet points of 'requirements' with no indication of what matters most.*
Genuine requirements	Only what is genuinely necessary. If a degree isn't required to do the job, don't list it.	*Credential inflation: MBA preferred,' 10+ years' experience, proven track record' without definition.*
How success is measured	Specific, observable outcomes for the first 6–12 months.	*Vague statements like 'exceeding expectations' or 'driving impact.'*
What the role offers	Specific, honest descriptions of what makes this role and organization worth joining.	*Generic employer brand language that sounds identical to every other JD on the platform.*
AI transparency	A single sentence explaining what AI is used in the process and what humans decide.	*Silence, which candidates increasingly interpret as concealment.*

Figure 6.4 - Competency-First framework

Using AI to Audit Your JD

AI is genuinely useful in the job description process, not to write the JD, but to audit it. Once you have a draft, use AI to ask **three specific** questions:

1. Does this JD contain language that could deter underrepresented candidates? Specific patterns to check: unnecessarily masculine language ('he will lead,' 'aggressive growth targets'), requirements that correlate with socioeconomic background (specific elite institution preferences, unpaid experience requirements), and physical or attendance requirements that could disadvantage disabled candidates or primary caregivers.

2. Does this JD contain requirements that are not genuinely necessary for the role? Degree requirements, years-of-experience thresholds, and specific credential

demands should all be tested against the question: could a person do this job excellently without meeting this requirement? If yes, the requirement is a barrier, not a filter.

3. Does this JD contain keywords or phrases that would be easy for an AI application generator to match without genuine capability? If the JD is full of industry buzzwords, it will generate applications full of the same buzzwords. Ask AI to identify the ten phrases most likely to be matched by a generator without requiring real experience and rewrite them in outcome-focused language.

The AI audit does not replace the human review. It surfaces issues for a human to make judgments about. That is the **Zone 2** model applied to job description development: AI identifies, human decides.

Restoring Candidate Trust

The candidate trust crisis is not primarily a communication problem. It is a design problem. Candidates do not distrust AI because it is mentioned in a disclaimer. They distrust it because they experience a process that feels automated, impersonal, and opaque. And, then they draw their own conclusions about the organization that designed it.

Restoring candidate trust requires two things: genuine transparency about what AI does in your process, and genuine human moments at the stages that matter most to candidates. Neither requires a large budget. Both require intentional design.

What Candidates Actually Want to Know

Candidate research consistently identifies **three questions** that candidates want answered about AI in recruitment processes. They rarely ask them directly, because they assume they will not get an honest answer. But **they assess you against the answers**, consciously or not, throughout the process.

- Is a human going to read my application? Not 'will a human be involved in the process'. And, will a human read my application before I am rejected?

- What is the AI actually assessing? Not a general statement about using AI 'to improve the candidate experience'. And, a specific description of what the AI does and what it does not do.

- Can I request a human review if I think the AI got it wrong? Not that it will necessarily change the outcome, but the option exists.

These questions are answerable. And answering them honestly, in plain English, in your job posting or candidate-facing communications, is a more powerful trust-building act than any amount of employer brand content.

The Transparency Statement

CANDIDATE-FACING AI TRANSPARENCY STATEMENT — TEMPLATE

We use AI in parts of our recruitment process. Here is what that means for you.

What AI does in our process:

- *Your application is acknowledged automatically and immediately.*
- *AI assists us in identifying applications that meet our minimum stated criteria.*
- *AI helps us track and analyze our recruitment pipeline for diversity and inclusion purposes.*

What humans decide in our process:

- *A named member of our talent acquisition team reviews every candidate who advances to our shortlist.*
- *Humans make all interview decisions and hiring decisions.*
- *Culture fit is assessed through structured human conversation, not algorithmic scoring.*

Your options:

- *If you believe your application has not been fairly assessed, you may request a human review by contacting [named person/email address].*
- *We will respond to all human review requests within [X] business days.*

We are committed to a fair process. If you have questions about how AI is used in our recruitment, please contact [named person].

Figure 6.5 - Transparency Statement Language

The Human Touchpoint Map

A human touchpoint map is a simple tool: a visual representation of every stage in your recruitment process, with a mark at each stage where a candidate currently has contact with a human being. In most organizations that have automated heavily, the number of genuine human touchpoints before the interview is zero.

The question the map asks is not "*how do we add more touchpoints*?" It is "*where would a human moment make the most difference to a candidate's experience of this process*?" The answer is almost always the same:

- the first communication **after application**, and
- the communication **around rejection**.

A candidate who receives an automated rejection after six weeks of silence has experienced a process that communicates institutional indifference. A candidate who receives a brief, specific, human communication at the same stage, even if the outcome is the same, has experienced something qualitatively different. The commitment of two minutes of a human's time is the difference between a candidate who tells their network to avoid your process and a candidate who applies again when the next role opens.

This is not a hypothetical cost. It is a calculable one.

> In 2014, Graeme Johnson, then Head of Resourcing at Virgin Media, set out to understand how rejected candidates perceived the brand. What he found stopped him cold. Of the roughly 123,000 candidates Virgin Media rejected each year, 18% were also paying customers. When Johnson's team cross-referenced rejection data against subscription cancellations, 6% of those rejected candidates canceled their subscriptions, approximately 7,500 cancellations per year. At a monthly subscription value of £50 ($60 USD), that translated to roughly £4.4 million ($5.4 million USD) in lost annual revenue.

And that figure, Johnson would later learn, was almost certainly an undercount. His own finance team estimated the true revenue impact was likely three to four times higher, once word-of-mouth was accounted for through candidates telling friends and family, who were also Virgin Media customers, about the experience. The CFO, presented with that math, approved the investment to fix it immediately.

The Virgin Media case matters for HR leaders in an AI-enabled environment for a specific reason: the problem was not that Virgin Media had too many candidates. It was that their process communicated institutional indifference at scale. The rejection was not the issue. The **experience of the rejection** was.

AI does not resolve this problem by default. An agentic workflow that screens 5,000 applicants overnight and routes 4,800 automated rejections by morning has not improved candidate experience. It has industrialized the exact dynamic that cost Virgin Media millions. Volume does not dilute the damage. It multiplies it.

What AI can do, deployed with intention, is make human communication possible at a scale that was previously unreachable. A recruiter cannot write 400 personalized rejection notes. But a recruiter can review, refine, and send 400 AI-drafted notes, notes that are specific, brief, and human in tone, in the time it once took to write five. The outcome is not efficiency. The **outcome is dignity at scale**. And **dignity**, as Virgin Media demonstrated, **has a revenue figure attached** to it.

Virgin Media's own analysis raised a follow-on question that reframes the entire conversation: what if a candidate experience so positive it earned new customers from the 82% of applicants who were not existing subscribers? Converting just 5% of that group represented an additional £3.3 million (roughly $4 million USD) in potential revenue. The hiring process, in other words, is not a cost center with a branding side effect. It is a market-facing function. It has always been. Most organizations simply have not run the numbers.

Your candidates are in your market. Treat them accordingly.

The financial services case study in this chapter measured this effect directly. A two-minute personalized audio message from a hiring manager to every shortlisted candidate recovered offer acceptance rates by 18 percentage points within two quarters. The cost was four minutes of hiring manager time per shortlisted candidate. The return was a pipeline that candidates wanted to stay in.

Measuring Candidate Trust

If you cannot measure candidate trust, you cannot manage it. The **leading indicators** are already available in most talent acquisition functions. And, they are not being tracked against the right question.

- A declining offer acceptance rate that is not explained by compensation factors is almost always a candidate experience signal. Track it against process changes, particularly the introduction of new AI tools. Offer acceptance rate.

- Candidates who begin the process and withdraw before completing it are telling you something. The stage at which withdrawals cluster is the stage at which the process is losing them. Application withdrawal rate.

- A simple post-process survey asking candidates (hired and rejected alike) whether they would recommend your recruitment experience to someone they know. Track this quarterly and against process changes. Candidate NPS.

- Candidates who were rejected and applied again at a later date are the strongest possible signal that your rejection experience was handled with enough respect to maintain the relationship. **Re-application rate.**

Cross Industry Case Study: Mid-Sized U.S. Financial Services

CROSS-INDUSTRY CASE STUDY

Organization: A mid-sized US financial services firm, approximately 6,800 employees, Talent Acquisition team of 8. **Function:** Talent acquisition across retail banking and corporate functions

The Problem:

Over an 18-month period following the introduction of an AI screening platform and an AI video interview tool, the organization's offer acceptance rate dropped from 84% to 61%. The TA team had assumed the decline was driven by compensation competitiveness. A candidate exit survey conducted in month 14 produced a different finding: 67% of candidates who declined offers cited the recruitment process itself as a factor, with the most common descriptor being "impersonal."

The AI video interview tool was identified as a particular friction point. Candidates were being asked to record responses to pre-set questions for an asynchronous AI assessment, with no human interaction until the shortlisting decision was communicated. The average time between video interview submission and outcome communication was 19 days. In that window, candidates received no human contact.

What They Did:

The TA team did not remove the AI tools. They redesigned the process around a single principle: every shortlisted candidate should encounter a genuine human voice before they receive a formal invitation to interview.

The intervention was deliberately low-tech. Within 24 hours of a candidate being shortlisted, the hiring manager recorded a two-minute audio message, not some generic welcome, but a specific reference to why the candidate's background was interesting for this role. The message was sent alongside the formal interview invitation.

The message took approximately four minutes of hiring manager time per candidate — two minutes to think, two minutes to record. It required no new technology. It required no additional training. It required only a commitment to the principle that a person entering your process deserves a human moment.

Outcome:

The offer acceptance rate recovered from 61% to 79% within two quarters. Candidate NPS improved by 34 points over the same period. Time-to-hire did not increase. And, the four-minute audio message was absorbed into the hiring manager's existing touchpoints without adding measurable drag to the overall process. Qualitative candidate feedback shifted: the most common descriptor of the process changed from "impersonal" to "they actually read my application."

The Lesson:

The trust deficit in AI-heavy recruitment processes is not solved by removing AI. It is solved by designing human moments at the stages that matter most to candidates. Four minutes of hiring manager time per shortlisted candidate produced an 18-percentage-point recovery in offer acceptance. That is one of the highest-ROI interventions in talent acquisition. It costs nothing but intention.

Figure 6.6 - Cross-Industry Case Study

GREY AREA DECISION FRAMEWORK

Grey Area: AI Video Interview Assessment

AI video interview tools that analyze facial expressions, tone of voice, or speech patterns to assess personality, culture fit, or suitability are among the most legally and ethically contested applications of AI in HR.

The core problem: these tools claim to measure something real about a candidate's suitability. The evidence that they do so reliably and without discriminatory bias is weak. The evidence that they disadvantage candidates from non-dominant cultural backgrounds, candidates with neurodivergent communication styles, and candidates for whom English is a second language is considerably stronger.

The test for any AI assessment tool in recruitment:

"Is this tool measuring something directly related to the requirements of this role, in a way I could explain and defend to a rejected candidate and, if necessary, an employment tribunal?"

Currently defensible uses:

- Structured competency question response analysis, whereby you assess the content of what a candidate says in response to defined competency questions, with human review of the output before any shortlisting decision
- Scheduling and logistics automation
- Communication and acknowledgment workflows

Currently indefensible uses:

- Facial expression analysis for personality or culture fit assessment
- Tone of voice or speech pattern scoring for suitability
- 'Culture fit' algorithmic scoring without a defined, validated, audited framework
- Any assessment that produces a score or ranking without a human reviewing the underlying basis

The Grey HR Zone in AI video assessment:

1. If you cannot explain to a rejected candidate exactly what the AI assessed and why that assessment is directly relevant to this role, do not use it.
2. If the vendor cannot provide a bias audit that covers the specific demographic groups in your candidate pool, do not trust the guarantee.
3. The absence of a complaint is not evidence that the tool is working fairly. It is evidence that rejected candidates do not know what to complain about.

CHAPTER CHECKLIST

☐ I have completed a **candidate experience audit** of our current recruitment process by mapping every stage from application to offer and identifying where candidates are eliminated without human review.

☐ I have mapped every stage of our process to the Automate / Collaborate / Protect model using the Hybrid Recruitment table in this chapter.

☐ I have identified at least one human touchpoint to introduce or strengthen and assigned a named owner responsible for it.

☐ I have reviewed our most frequently used job descriptions for AI-screening bias using the JD audit prompts in this chapter.

☐ I have drafted or scheduled the development of a candidate-facing transparency statement about how AI is used in our recruitment process.

PROMPT LIBRARY & WORKBOOK EXERCISES

JD AUDIT PROMPT

[Analyze this job description for:

1. *language that could deter underrepresented candidates, including unnecessarily gendered language, socioeconomic proxies, and credential inflation.*

2. *Requirements that are not genuinely necessary for the role and test each requirement against the question: could a person be excellent in this role without meeting it?(3) keywords or phrases that would be easy for an AI application generator to match without genuine capability. For each issue identified, suggest a specific rewrite in competency-first, outcome-focused language.]*

CANDIDATE TRANSPARENCY PROMPT

[Draft a candidate-facing statement explaining how AI is used in our recruitment process. It should be honest and specific about what AI does, what humans decide, and how candidates can request human review. Tone: direct and reassuring, not defensive or legalistic. Maximum 200 words. It should appear in our job postings and on our careers page.]

WORKBOOK EXERCISE

Map your current recruitment process from application submission to offer accepted. At each stage, mark: AI-only decision, human-only decision, or AI surfaces/human decides. Count the number of stages where a candidate could be eliminated without any human ever seeing their application. That number is your starting reform point. Then identify the single stage where introducing or strengthening a human moment would have the greatest impact on candidate experience.

TEAM EXERCISE: THE SHADOW APPLICANT TEST

Have two members of your team apply for one of your currently open roles as shadow candidates. One should write their application authentically, describing their experience in plain, human language without keyword optimization. The other should use an AI application generator to produce a keyword-optimized application. Run both through your screening process. Debrief on which application the AI screener preferred, and what that tells you about what your process is actually selecting for.

SKEPTIC'S CORNER FAQ

Q: "We need AI to manage application volume and we simply can't screen manually."

A: Nobody is suggesting you screen 340 applications manually. The hybrid model in this chapter uses AI for administrative triage and human judgment for shortlisting decisions. The question is not AI or human. It is where exactly the human comes back into the process. And, whether that re-entry point is before or after the candidates most likely to be genuinely excellent have been filtered out. Tom's three-week manual re-screen costs more time than a well-designed hybrid process would have. Volume is not the problem. Process design is.

Q: "Candidates who use AI to write applications are just being efficient."

A: Some are. Others are submitting applications for roles they are fundamentally unsuited for, at scale, because the cost of applying is now effectively zero. Both groups are generating volume that your screening process needs to handle. The answer is not to penalize AI-generated applications. You cannot reliably detect them, and the attempt creates its own bias risks. The answer is to design your screening process to assess demonstrated capability rather than keyword density, and your JD to attract human thinking rather than algorithmic matching.

Q: "We can't afford to add human touchpoints and our TA team is already overstretched."

The candidate experience audit often shows the same pattern. TA teams spend time on low-value administrative tasks that AI could handle. At the same time, high-value human moments like shortlisting review, candidate communication, and offer conversations are rushed or skipped. This is not about adding work. It is about redistributing it. In the Canadian case study, four minutes of hiring manager time per shortlisted candidate replaced time previously spent on automated emails that no one read.

What Tom Does Differently Now

Tom does not stop using his AI screener. He changes what he asks it to do. He configures it to surface candidates and not to select them. The output of the screen is a candidate pool for human review, not a shortlist. A named member of his team reviews every candidate the AI surfaces before any shortlisting decision is made. **The AI finds. The human decides.**

He also rewrites the job descriptions for his highest-volume roles using the **Competency-First framework**. He asks AI to audit each one for credential inflation, gendered language, and keyword-gaming risk. **He cuts requirements that do not genuinely relate to the role**. The applications he receives are fewer. The shortlists his team produces are better.

And then he adds one human touchpoint that costs four minutes and changes everything: a brief, specific, personal message from the hiring manager to every shortlisted candidate, sent the same day they're selected, before the formal interview invitation. Four minutes is all it takes to ***start acting like a workplace of choice***.

The candidates who receive that message know they are being considered by a person, not processed by a system. And the difference between those two experiences, between being considered and being processed, is the difference between a candidate who accepts your offer and a candidate who doesn't. That is what it means to design human judgment back into a polluted pipeline.

AI Adoption – *62% of organizations are now using AI somewhere in their operations. Only 39% have it adopted directly within HR. The gap between what the business is doing and what HR is governing is already open.*

~ SHRM, State of AI in HR, 2026

Chapter 7: What AI Makes Possible for L&D | Organizational Development | RBPs

Before You Build the Course . . . The Right Intervention for the Right Problem

CASE IN POINT

Carlos is the VP of Operations at a regional logistics company with 1,800 employees across six distribution centers. He is decisive, data-driven, and genuinely invested in his people. He is also, by his own admission, a chronic over-trainer.

Figure 7.1 - Carlos' Story

In the last twelve months, Carlos has submitted fourteen training requests to the L&D function. Supervisor communication skills. New hire onboarding refreshers. Inventory accuracy procedures. Forklift safety recertification. Pick rate performance.

On and on. Each request arrived with the same diagnosis attached: people need training.

His HRBP, a sharp L&D specialist named Priya who has been quietly watching the pattern, finally asked him a question he was not expecting.

"Carlos, how many of those fourteen training programs produced a measurable change in the performance you were trying to fix?"

He did not answer immediately. Because the honest answer was three. Maybe four.

Priya did not say he was wrong to request training. She said something more useful: she was not sure training had been the right intervention for most of them. And she had a framework she wanted to walk him through before he submitted the next one.

Carlos is not unusual. He is the majority. Across industries, functions, and management levels, training is the default response to performance problems not because managers are lazy or uninformed, but because nobody has ever given them a better first question to ask.

***

This chapter gives them that question. And it gives HRBPs and L&D/OD specialists the diagnostic infrastructure to make sure the right answer follows.

The $400 Billion Model That Isn't Working

In 1990, *Peter Senge* published ***The Fifth Discipline*** and introduced the world to the Learning Organization. A concept so compelling and so correct that it has been cited, celebrated, and largely unrealized for thirty-five years. That same year, *Geary Rummler* and *Alan Brache* published Improving Performance and told us exactly why training fails when the system around the performer is broken. Thomas Gilbert had been making the same argument since 1978. The frameworks were right. The diagnostic tools existed. The infrastructure to apply them at the individual, team, and department level across an entire organization did not.

That infrastructure now exists. It is called AI. And it changes everything the past promised but could not often deliver.

The traditional L&D model that survived those thirty-five years was built on a simple and deeply flawed premise: if you push enough content at enough people, performance will improve. Design a course for a job category. Roll it out to everyone in that category. Measure completions. Assume impact. The model was efficient in a narrow sense. It consumed HR budget and produced learning records. What it reliably did not produce was changed performance.

The scale of the investment makes the failure more striking, not less. Organizations collectively spend $400 billion annually on training, content libraries, L&D technology, and learning consultants. *Josh Bersin's* research across hundreds of companies found that 74% of organizations still report they cannot keep pace with their own demand for new skills. If three-quarters of organizations are falling behind after thirty-five years and $400 billion, the investment is not the issue. The model is.

The Learning Organization Senge described was not a training program. It was a system. It required feedback loops, shared mental models, systems thinking, and continuous diagnostic capability at every level of the organization. Most organizations read The Fifth Discipline, ran a workshop, and called it done. The concept demanded infrastructure they did not have and could not build with the tools available to them at the time.

That is no longer the constraint.

The Performance Problem Nobody Is Governing

Before we discuss what AI can do for L&D, we need to talk about what is already happening without L&D's knowledge or involvement. Because in most organizations, a shadow training ecosystem is operating in plain sight. And HR rarely has a handle on it.

Call it ghost training. A sales team is underperforming against quota. The sales director pulls the top performers into a room and has them show the rest of the team how they do it. A new hire joins a busy operations department. There is no formal onboarding program, so their manager assigns them a buddy and tells them to shadow for two weeks. A customer service team is struggling with a new system. The team lead, who figured it out first, spends three hours walking everyone else through it one by one.

None of this is designed. None of it is evaluated. None of it is coordinated with L&D. And none of it is necessarily wrong. Peer learning, mentoring, and on-the-job experience are legitimate and often highly effective development interventions. The problem is not that it is happening. The problem is that it is happening as a default, driven by operational urgency, with no diagnostic foundation and no governance framework. Nobody asked whether training was the right intervention. Nobody asked what was actually causing the performance gap. The manager saw a problem, reached for the most visible solution available, and moved on.

This is not a failure of management intent. It is a failure of infrastructure. Managers default to training, formal or informal, because nobody has ever given them a better tool or a different question to ask.

The HRBP who should be catching this is often not equipped to catch it. Diagnosing skill versus will, distinguishing a capability gap from a motivation problem, identifying when a performance issue is rooted in the environment rather than the individual, these require a fluency in human performance technology that most HRBPs have not been trained in. They can spot a symptom. They cannot always prescribe the right intervention. And without a diagnostic framework, they often arrive at the same conclusion the manager already reached: send them to training.

Six Sigma and continuous improvement disciplines got closest to asking the right question in large organizations, is this a process problem rather than a people problem? But they never fully crossed into human performance territory, and they never scaled to mid-market or smaller organizations where the ghost training problem is most acute and most ungoverned.

The result is an L&D function that is simultaneously over-invested in content and under-invested in diagnosis. More courses. More platforms. More content libraries. And a 74% failure rate on keeping pace with skills demand.

The frameworks to diagnose and fix this have existed for decades. What was missing was the infrastructure to apply them consistently, at scale, and in real time.

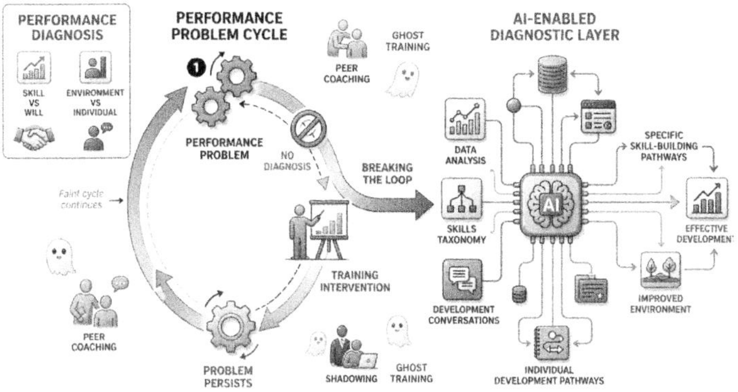

Figure 7.2 – Performance Problem Cycle

What Gilbert Actually Said

Thomas Gilbert's **Behavior Engineering Model**, published in *Human Competence: Engineering Worthy Performance* in **1978**, made an argument that the training industry has spent nearly five decades politely ignoring: most performance problems are not caused by a lack of knowledge or skill. They are caused by the environment in which the performer works.

Gilbert's model identifies six factors that influence human performance, organized across two dimensions. The environmental factors, information, resources, and incentives, sit outside the performer and account for the majority of performance variance. The individual factors, knowledge, capacity, and motivation, sit inside the performer and are the factors training can actually address. Gilbert's central insight was that organizations almost always intervene at the individual level when the problem almost always lives at the environmental level.

A performer who lacks clear expectations, adequate tools, and meaningful feedback will underperform regardless of how much training they receive. A performer who has all three will largely self-correct without any training at all. The intervention has to match the root cause. Training is the right answer only when the root cause is a genuine knowledge or skill deficit. That is one quadrant of four. We'll explore all four quadrants shortly.

The Three Levels and Where the Gap Lives

Rummler and *Brache* gave practitioners the architecture to understand why this matters at an organizational scale. Their three-level framework, Organization, Process, and Job/Performer, maps exactly where performance problems originate and where interventions need to land.

Organizational Development has historically operated at the first two levels. Culture, structure, strategy, change readiness, and organizational network analysis are Organization and Process level work. It is important work. It is also insufficient on its own, because most performance gaps do not originate at the organizational level. They originate at the Job/Performer level, in the daily interaction between a specific person, a specific role, and a specific set of environmental conditions.

This is the level that has been chronically under-resourced and under-governed. Not because the field did not know it mattered. *Rummler* and *Brache* named it in 1990. *Gilbert* named it in 1978. But applying a rigorous performance diagnostic at the individual and team level, across an entire organization, for every performance gap that surfaced, required a depth of human expertise and a volume of practitioner time that most organizations simply could not resource.

Until now.

Two Questions That Should Precede Every Training Decision

The Performance Analysis Quadrant (PAQ) is not a new idea. It has been in the HPT practitioner's toolkit for decades, grounded in the work of *Gilbert* and formalized within *Rummler* and *Brache's* Job/Performer level framework. It is also one of the most underused diagnostic tools in the field. Not because it is complicated. Because nobody built the infrastructure to make it a standard part of how performance problems get solved.

The PAQ starts with two questions.

1. **Does the employee have adequate knowledge (SKILL) to perform the job?**

2. **Does the employee have the motivation and desire (WILL) to perform the job?**

Rate each on a scale of one to five. Plot the answers. The result places the performance gap in one of four quadrants, each with a different root cause and a different intervention. It takes fifteen minutes. It changes the entire conversation.

Here is what each quadrant tells you and what it means for the intervention.

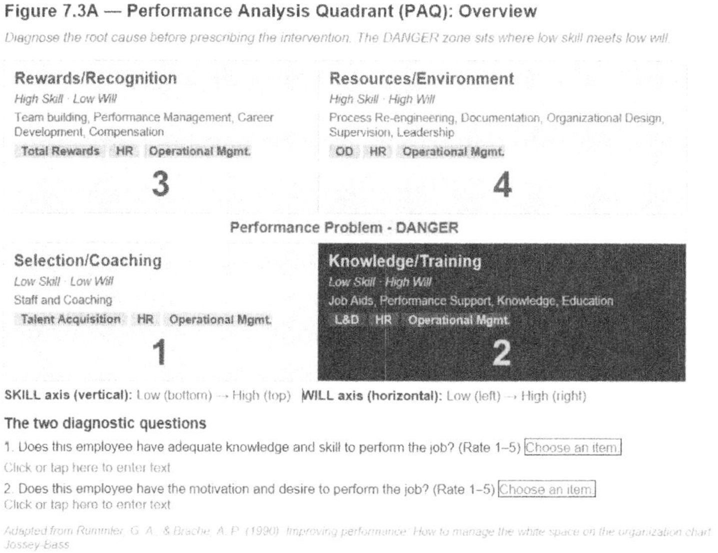

Figure 7.3A – Performance Analysis Quadrant (PAQ)

Quadrant #1: The Selection/Coaching Problem

The employee lacks both the knowledge and skill to perform the job and the motivation to develop them. This gap is fundamental. The person may be misplaced in the role, wrongly hired for it, or promoted beyond their current capability and their willingness to close the gap.

This is the most difficult conversation in people management. And the one most often avoided by defaulting to training. An organization that sends a fundamentally misplaced employee through a development program is not investing in their growth. It is delaying a decision that should have been made earlier and will have to be made eventually, at great cost to the employee, the team, and the organization.

The intervention for Quadrant #1 is a frank assessment of fit. That might mean a role transfer to a position that better matches the employee's strengths and motivations. It might mean a structured performance improvement process with honest expectations. It might mean a separation conversation. None of those interventions is easy. None of them is a training course.

Quadrant #2: The Knowledge/Training Problem

The employee wants to perform. They have the motivation, the attitude, and the desire. What they lack is the knowledge or skill to do the job at the required level. This is the only quadrant where training is the right primary intervention.

This is the new hire who is genuinely trying and genuinely learning but needs structured development to close the gap between their current capability and the role requirement. This is the experienced employee moving into a new function who has the will but not yet the technical knowledge. This is the team being asked to operate new technology with no transition support and no time to self-discover.

Training solves Quadrant #2 problems. It does not solve Quadrant #1, #3, or #4 problems. And the evidence suggests that Quadrant #2 is the least common root cause of performance gaps, not the most common, which is precisely what the traditional L&D model was built to assume.

Quadrant #3: The Rewards/Recognition (Motivation) Problem

The employee knows how to do the job. They have the skills, the knowledge, and the experience. But they are not performing. The gap is not capability. It is consequence, recognition, or incentive alignment.

This is the sales representative who consistently closes deals below her capability because the commission structure rewards volume over margin and she has quietly optimized for her own time. This is the tenured operations employee who stopped raising process improvement ideas after the last three went nowhere. This is the high-potential manager who has disengaged because nobody has had a development conversation with him in two years.

Training will not fix any of these situations. More learning content directed at a motivation problem is not just ineffective. It is insulting to the employee who already knows how to do the job and is watching the organization treat a management or culture failure as a skills gap.

The intervention for Quadrant #3 is a conversation about consequences, recognition, and environment. It is a manager coaching discussion. It is a look at what the role actually rewards and whether that reward structure produces the behavior the organization says it wants. It is human work. AI can surface the signal. The human has to act on it.

Quadrant #4: The Resources, Process, and Environment Problem

The employee knows how to do the job and wants to do it. But performance is still unsatisfactory. The problem is not inside the employee. It is in the system around them.

This is the customer service representative who cannot resolve calls at first contact because the **internal knowledge base** is three years out of date and nobody has budget to update it. This is the new hire who is struggling through week four not because they lack ability but because the **onboarding buddy** they were assigned is **overloaded** and the **process documentation** is **inconsistent**. This is the manufacturing team whose error rate spiked after a system upgrade that came with **no transition support** and no **updated job aids**.

> Training will not fix any of these situations either. Sending someone to a course when the problem is a broken process, an inadequate tool, or an unsupportive environment is one of the most expensive ways an organization can demonstrate that it does not understand its own performance problems.

The intervention for Quadrant #4 is process redesign, resource allocation, or environmental adjustment. It is the kind of diagnosis that Six Sigma got partially right and that OD has addressed at the organizational level. AI can now help surface it at the individual and team level, through performance data patterns, workflow analysis, and the micro-level diagnostic that was previously too resource-intensive to run consistently.

Figure 7.3B — PAQ: Causes and Interventions Reference

Red indicators identify causes of performance gaps. Blue indicators identify HR and management interventions. Green indicators identify L&D interventions.

Quadrant	Causes of the performance gap	Interventions
1 — Selection/Coaching *Low Skill · Low Will*	☐ Understaffed ☐ Individual capacity limitations ☐ Staffing misalignment ☐ Health and wellness factors	☐ Staffing review and adjustment ☐ Structured coaching protocol ☐ Probationary support plan ☐ Role transfer or separation if coaching does not produce movement
2 — Knowledge/Training *Low Skill · High Will*	☐ Insufficient product or role knowledge ☐ Lack of time and know-how ☐ Inadequate onboarding or orientation ☐ No access to information systems or documentation	☐ Job aids and performance support tools ☐ Structured learning and training program ☐ Knowledge base and documentation access ☐ Mentorship and on-the-job coaching ☐ Job and work design support
3 — Rewards/Recognition *High Skill · Low Will*	☐ Lack of recognition and acknowledgment ☐ Misaligned incentive structure ☐ Insufficient or inequitable compensation ☐ No clear career development pathway ☐ Ineffective performance appraisal process	☐ Recognition and rewards program review ☐ Compensation and incentive restructuring ☐ Team building and engagement initiatives ☐ Career development planning and pathing ☐ Performance management redesign ☐ Leadership coaching for managers
4 — Resources/Environment *High Skill · High Will*	☐ Inadequate physical resources or tools ☐ Poor data access or information systems ☐ Lack of support or feedback mechanisms ☐ Broken or inefficient processes ☐ Cultural barriers or organizational dysfunction ☐ Ineffective supervision or leadership	☐ Process re-engineering and workflow redesign ☐ Tool and resource allocation review ☐ Organizational design adjustment ☐ Supervision and leadership development ☐ Culture change initiative ☐ Documentation and information systems upgrade ☐ Feedback and support structure redesign

Legend:
• Causes of performance gap • HR and management interventions • L&D interventions

Adapted from Rummler, G. A., & Brache, A. P. (1990). Improving performance: How to manage the white space on the organization chart. Jossey-Bass.

Figure 7.3B – Performance Analysis Quadrant (PAQ) Causes & Interventions Reference

A Note Before You Diagnose

The PAQ works. It has always worked. But it works best when one condition is met: the performance expectation for the role is clearly defined before the diagnostic begins.

As performance consultant Joe Dunlap observes, you cannot diagnose a performance gap against a standard that does not exist. If the manager submitting the training request cannot articulate what good performance actually looks like in that role, under real conditions, the PAQ questions will produce a confident quadrant placement that may be built on an unstable foundation.

This does not mean you wait for perfect clarity before running the diagnostic. In many organizations, that clarity will never arrive on its own. What it means is that the PAQ conversation may need to start one step earlier than you expect.

If the manager cannot answer the skill and will questions with confidence, that is itself diagnostic information. The first performance gap you may need to address is not the employee's. It is the absence of a defined standard. That is a Quadrant 4 finding before the Quadrant 4 finding. The environment is missing something the performer needs to succeed, and in this case, that something is a clear expectation of what success looks like.

AI can help here too. An AI-assisted diagnostic conversation can prompt the manager to articulate the performance standard before the quadrant is assigned. The output becomes more useful. The diagnosis becomes more defensible. And HR has done something more valuable than routing a training request correctly. It has surfaced a performance architecture gap the organization did not know it had.

Why This Changes Everything

The PAQ has been available to practitioners for decades. The reason it never fully penetrated operational management is not that business line managers resist diagnostic thinking. It is that applying it consistently required a knowledgeable performance consultant in a one-on-one conversation with a manager, for every performance gap, across every team, in every department. That is not a realistic resource model for any L&D function operating with a team of three and a compliance training backlog.

One constraint worth naming explicitly. AI-assisted diagnosis is only as good as the information it is working from. If performance expectations are inconsistently defined, workflows are undocumented, or the data reflects organizational variability rather than clear standards, the diagnostic output will reflect that variability rather than correct it. The PAQ questions sharpen the conversation. They do not manufacture clarity that does not exist in the organization. Where the inputs are solid, AI makes the diagnostic faster and more consistent. Where they are not, the diagnostic conversation becomes the first step toward building them.

This is not AI replacing a specialist most organizations never had. Formal human performance consulting has historically lived in a narrow corner of the field. A select group of OD practitioners. A small cadre of instructional designers and developers who found their way to performance improvement frameworks through ISPI or advanced study. The occasional HRBP fluent enough in skill versus will to push back on a training request that should have been a coaching conversation. Even at the professional conferences dedicated to human performance technology, the room is filled mostly with OD consultants and academics. The corporate trainer, the L&D coordinator, the HRBP fielding requests from a sales director who has already decided the answer IS training, they were never in that room. Nobody brought the frameworks to them.

AI changes who can do this work. PAQ logic in an AI-assisted tool does not require HPT certification or an OD background. It requires two questions and the discipline to ask them before opening the course catalog. It also requires the confidence to tell an operations manager that the request is not the need. That capability can now exist in any L&D team that is paying attention. This is not an efficiency gain. It is a structural shift.

2026 and the L&D Profession Gets What It Always Needed

Peter Senge did not fail. The Learning Organization did not fail. *Gilbert* did not fail. *Rummler* and *Brache* did not fail. The frameworks were right. The tools existed. What failed was the assumption that an idea powerful enough to be true would somehow find its own infrastructure.

It did not. And for decades, L&D practitioners carried frameworks they could not fully deploy, asked questions the organization did not have time to answer, and watched performance problems get addressed with training courses that were never going to fix them. Not because the practitioners were wrong. Because the infrastructure to do the work correctly, at scale, at the individual level, consistently and affordably across an entire organization, simply did not exist.

That infrastructure now exists. And it changes three things simultaneously that have never changed together before.

First: The Diagnostic Becomes Scalable

The **Performance Analysis Quadrant** works. It has always worked. The problem was never the tool. The problem was that applying it consistently required a skilled performance consultant in a one-on-one conversation with a manager, for every performance gap, across every team, in every department. That is not a realistic resource model for any L&D function operating with a team of three and a compliance training backlog.

AI changes the economics of diagnosis entirely. The two PAQ questions become a structured AI-assisted diagnostic conversation that any manager, any HRBP, or any L&D practitioner can run before a training request is ever submitted. The ghost training problem gets a first line of defense. The operations manager who has always defaulted to sending people to training now has a fifteen-minute diagnostic process that tells them whether training is actually what they need.

And critically, this does not require the organization to have an HPT specialist on staff. It does not require an ISPI membership or an OD background or a CPTD with a performance improvement emphasis. It requires a practitioner willing to ask two questions before accepting a training request at face value. AI makes that practitioner possible in every organization, not just the ones that have been lucky enough to have the right person in the right role.

Second: The Practitioner Is Freed for the Work That Matters

Ask any trainer, instructional designer, or L&D coordinator what they actually spend their time on. The answer is almost never performance consulting. It is content development. Course maintenance. LMS administration. Compliance tracking. Onboarding coordination. Reporting. The work that fills the calendar is not the work that changes performance. It is the work that keeps the function running.

AI does not eliminate that work. It absorbs it. Content generation, curation, and pathway design, the activities that consume the majority of most L&D practitioners' productive time, become AI-assisted tasks that take a fraction of the time they previously required. Agentic workflows handle compliance tracking, certification expiration, learning reminders, and completion reporting without human intervention. The administrative layer that has always competed with the strategic layer for practitioner time moves off the human plate entirely.

What remains is the work that AI cannot do. The diagnostic conversation with the operations manager who thinks her team needs training but actually has a motivation and consequence problem sitting squarely in Quadrant 3. The coaching discussion with the HRBP who has been accepting training requests without asking whether training is the right intervention. The performance consulting relationship with the business line leader who has never had an L&D partner who asked the right questions before opening the course catalog.

This is the work the L&D profession has always said it wanted to do. AI is not taking it away. It is finally clearing the path to do it.

Third: The Micro and the Macro Finally Connect

Organizational Development has always operated at the system level. Culture, structure, strategy, network analysis, change readiness. These are organizational and process level interventions. They are important. They are also incomplete without a corresponding capability at the individual and team level.

The Learning Organization Senge described required feedback loops at every level of the organization, not just the top. The skills-based organization the CEO is asking Diane to build requires performance data at the individual level, not just the workforce planning level. The change readiness assessment OD produces is only actionable if the practitioners who need to respond to it can diagnose and address performance gaps at the team and department level.

AI connects these levels for the first time. Macro-level OD findings translate into micro-level intervention recommendations. Micro-level PAQ diagnostics feed upward into the organizational performance picture that OD needs to work with. The three *Rummler-Brache* levels, Organization, Process, and Job/Performer, finally talk to each other in real time, with AI as the connective tissue between them.

This is what the Learning Organization always required and never had. Not more training. Not a better LMS. A diagnostic and feedback infrastructure that operates at every level simultaneously and connects the insights from each level to the interventions at the others.

The New Professional Identity

What emerges from these three shifts is not a retrained trainer or instructional designer/developer. It is a genuinely different professional.

The human performance consultant operating in an AI-enabled L&D function is not doing what trainers and instructional designers have historically done. They are doing what OD practitioners and the small cadre of HPT specialists have done, at the individual and team level, inside the organization, as a standard part of how performance problems get diagnosed and addressed. The work that was once confined to a narrow specialist community visible mainly at ISPI conferences is now accessible to every L&D practitioner willing to pick up the tools.

They are the person the operations manager calls before sending someone to training. They are the practitioner who sits with the HRBP and runs a PAQ analysis on the performance gaps surfaced in the last engagement survey. They are the professional who takes the organizational network analysis OD just produced and translates it into individual development conversations with the informal influencers it identified.

They are not a new job title. They are a new capability, built from the same people who were already in the L&D function, equipped with tools that finally match the sophistication of the frameworks the field has always known were right.

Diane's team of three does not need to become five. It needs to become different. AI handles the content. The practitioners handle the diagnosis. And the organization finally gets what it has been paying for and rarely receiving: performance change that lasts.

This is not a career adjustment for the L&D practitioner. It is a career redefinition. And it is the most significant shift the profession has seen since the first learning management system convinced an entire generation that technology plus content equaled performance.

It does not. It never did. But this time, with the right diagnostic infrastructure and the right human judgment at the center of it, it finally can.

Cross-Industry Case Study: Mid-Sized Manufacturing Company

CROSS-INDUSTRY CASE STUDY

Organization: A mid-sized US manufacturing company, approximately 2,400 employees across three production facilities

Function: Human Resources and Operations, Learning & Development

The Problem:

Over an eighteen-month period, the VP of Operations at a Midwest-based packaging manufacturer had submitted twenty-two training requests to the company's two-person L&D function. The requests spanned supervisor communication skills, safety compliance refreshers, equipment operation procedures, quality control standards, and new hire onboarding reinforcement. The L&D team built and delivered against every request. The training completion rates were strong. The performance problems persisted.

The HRBP assigned to the operations function had been watching the pattern quietly. Turnover in two of the three facilities was running at 34% annually. Quality defect rates had not moved despite three rounds of quality control training in fourteen months. A safety incident in the second quarter had prompted an immediate refresher course. A second incident six weeks later suggested the course had not addressed the root cause. Nobody had asked what the root cause actually was.

The HRBP requested a meeting with the VP of Operations and arrived with one question: before the next training request, could they spend thirty minutes diagnosing why the performance gap existed rather than what content might close it? The VP agreed. Reluctantly.

What They Did:

The HRBP, with the help of one of the L&D Specialists, introduced the Performance Analysis Quadrant as a structured intake process for every performance gap before a training solution was commissioned. Using an AI-assisted diagnostic tool built on PAQ logic, the team worked through the twenty-two previous training requests retrospectively and then applied the same process to the four active performance gaps currently on the VP's radar.

The retrospective analysis was uncomfortable. Of the twenty-two previous requests, nine were genuine Quadrant 2 Knowledge/Training problems. Training had been the right answer for those nine. Seven were Quadrant 3 Rewards/Recognition problems tracing to a supervisor whose management style was creating disengagement, a compensation structure that rewarded individual output over team quality, and a promotion process employees perceived as opaque. Four were Quadrant 4 Resources/Environment problems. The safety incidents were the most significant finding: both occurred on the same shift, on the same line, under the same supervisory configuration. The root cause was not a knowledge deficit. It was a staffing ratio that created time pressure at the exact moment the safety protocol required the most attention. Two were Quadrant 1 Selection/Coaching problems that had been quietly avoided for over a year.

Outcome

The organization implemented a PAQ-based intake process for all performance-related requests across the three facilities. Within two quarters, the quality defect rate dropped by 22%. The safety incident rate dropped to zero for the remainder of the year. Turnover at the affected facility dropped from 34% to 21%. Training spend decreased by 31% in the twelve months following implementation. The L&D function did not shrink. It redirected.

The Lesson:

The performance problem was never the training. The training was the answer to a question nobody had stopped to ask. Twenty-two requests. Eighteen months. A safety incident that happened twice. None of it required more content. All of it required a better first question.

Figure 7.4 - Cross-Industry Case Study

Grey Area: The PAQ as the Transparency Test

There is a version of AI-assisted performance diagnosis that belongs in Zone 2 and a version that belongs in Zone 3. The difference is not the technology. It is the transparency.

A Zone 2 application looks like this: a manager and an L&D practitioner sit together, run the two PAQ questions against a specific performance gap, and use an AI tool to structure the diagnostic conversation and generate intervention recommendations. The employee knows a performance conversation is happening. The data being used is observable and discussable. The human makes the judgment. The AI supports the process.

A Zone 3 application looks like this: an AI tool analyzes an employee's communication patterns, meeting participation frequency, Slack activity, and document collaboration behavior, assigns them an inferred performance or potential score, and surfaces that score in a succession planning or development decision without the employee's knowledge or consent. The diagnostic is invisible. The employee cannot challenge it. The intervention recommendation arrives from a process they did not know was running.

The PAQ transparency test is simple. Could you sit across from the employee whose performance gap you are diagnosing and describe exactly what data you used, what questions you asked, and what the AI recommended? If yes, you are in Zone 2. If the answer is no, or if the honest answer is that the employee does not know the diagnostic is happening at all, you are in Zone 3. Stop. Redesign the process with the employee's awareness and participation before proceeding.

Performance consulting has always required trust between the practitioner and the person whose performance is being discussed. AI-assisted performance diagnosis is no different. The tool changes. The obligation does not.

CHAPTER CHECKLIST

☐ I have mapped my last quarter's training requests against the four PAQ quadrants and identified how many were genuine Quadrant 2 problems versus problems that belonged in a different quadrant entirely.

☐ I have identified the ghost training activity currently operating in my organization, peer coaching, buddy systems, informal manager-delivered training, and established a first-line governance response.

☐ I have assessed whether my HRBPs/L&D/OD staff have the PAQ fluency to push back on a training request that belongs in a different quadrant and identified what support they need to build that capability.

☐ I have shared the quadrant-to-HR-function map with my HR leadership team and discussed which quadrant our organization most frequently misdiagnoses as a training problem.

☐ I have run the **PAQ Diagnostic Prompt** on at least one current performance problem and brought the output into a conversation with the requesting manager before recommending an intervention.

PROMPT LIBRARY & WORKBOOK EXERCISES

PAQ DIAGNOSTIC PROMPT

Use this prompt before accepting any training request. Run it in your AI tool of choice. Bring the output into your conversation with the requesting manager.

"A manager has submitted a request for training to address a performance gap on their team. Before recommending a training intervention, I need to run a performance diagnostic. Here is what I know about the situation:

- Role: [describe the role]
- Performance gap: [describe what the employee is not

doing or not doing well]

- Tenure in role: [how long the employee has been in this position]

- Context: [any relevant information about the team, the environment, the manager, recent changes, or workload]

Using the Performance Analysis Quadrant framework, answer the following:

- First, based on the information provided, rate the employee's likely knowledge and skill level for this role on a scale of one to ten and explain your reasoning.

- Second, based on the information provided, rate the employee's likely motivation and desire to perform on a scale of one to ten and explain your reasoning.

- Third, identify which PAQ quadrant this performance gap most likely belongs to: Quadrant 1 Selection/Coaching, Quadrant 2 Knowledge/Training, Quadrant 3 Rewards/Recognition, or Quadrant 4 Resources/Environment

- Fourth, recommend three specific interventions appropriate for this quadrant. If the quadrant is not Quadrant 2, explain why a training solution is unlikely to resolve the performance gap.

- Fifth, identify any additional information I should gather from the manager before confirming the diagnosis and recommending an intervention."

How to use this prompt:

Run it before the follow-up conversation with the requesting manager. Use the AI output as your diagnostic starting point, not your final answer. The AI generates the hypothesis. You test it in the human conversation. If the manager's answers shift the quadrant, update the recommendation. The prompt does not replace the conversation. It prepares you for it.

WORKBOOK EXERCISE

Pull your last ten training requests, formal or informal, from the past quarter. For each one, apply the two PAQ questions retrospectively. Rate the employee's likely knowledge and skill level and their likely motivation and desire to perform. Plot each request in a quadrant. What percentage landed in Quadrant 2? What should the others have been? That gap is your organization's performance misdiagnosis rate. It is also your starting point for a different kind of conversation with your operational leaders.

TEAM EXERCISE: THE PAQ LIVE DIAGNOSTIC

Bring a current, real performance problem into your next HR leadership team meeting. Do not tell the group what intervention has already been requested or implemented. Run the two PAQ questions as a group. Rate skill and motivation independently. Plot the quadrant together. Then reveal what intervention was actually deployed. If the quadrant and the intervention do not match, that is your case study. Use it to build the argument internally for a PAQ-based intake process before the next training request is accepted.

SKEPTIC'S CORNER FAQ

Q: "We don't have time for a performance diagnostic every time a manager submits a training request. That is why we just build the course."

A: That is exactly the thinking that produces the $400 billion investment with a 74% failure rate. The diagnostic does not take long. Two questions. Fifteen minutes. A structured AI-assisted prompt that any L&D practitioner can run before the follow-up conversation with the requesting manager. The question is not whether you have time for the diagnostic. It is whether you have time to build a course that does not fix the problem and then wonder why the performance gap is still there six months later.

The ghost training ecosystem in your organization exists partly because operations and sales stopped believing L&D could solve their performance problems. Not because the problems were unsolvable. Because the solutions were mismatched to the root cause. A team that keeps receiving training for a motivation problem, or a resources problem, or a selection problem, eventually stops requesting training and starts solving it themselves. Informally. Without governance. Without evaluation. Without L&D.

In many cases, the informal learning is not just ungoverned. It is filling a gap that a defined workflow or performance standard should have occupied. People are not defaulting to peer coaching and buddy systems because they prefer it. They are reconstructing the work in the absence of a clear standard for how the work is supposed to be done. The ghost training ecosystem is a symptom. The missing performance architecture is the disease.

The PAQ diagnostic is the first step toward building that architecture, and it starts with fifteen minutes before the next course request is accepted.

Q: "Our managers will never sit still for a fifteen-minute diagnostic conversation. They just want a solution."

A: The fifteen minutes is not an additional meeting. It is a different conversation inside the meeting that was already happening. The manager who submits a training request is already talking to someone in HR or L&D. The PAQ changes what that conversation asks, not how long it takes. Two questions. A rating. A quadrant. And then a recommendation that actually matches the problem. That is not more time. That is better time. The manager who keeps getting training solutions for motivation problems will eventually stop calling HR at all. The PAQ is what keeps them in the conversation.

Q: "Isn't this just common sense? Good managers already know whether something is a training problem or a motivation problem."

A: Some do. The evidence suggests most do not . . . not consistently, not under operational pressure, and not without a structured framework to anchor the diagnosis. Across industries and management levels, training is the default response to performance problems regardless of root cause. Not because managers are careless. Because nobody gave them a better first question. Common sense is not consistently applied when the deadline is Friday and the VP is asking why the numbers are down. The PAQ makes the right question structural rather than dependent on any individual manager's diagnostic instinct on any given day. AI makes it scalable. That is the difference between a good idea and a governance system.

What Carlos Discovers

Carlos does not wait for Priya to follow up. The morning after their conversation he pulls up his training request log from the last twelve months and sits with it for an hour.

Fourteen requests. He runs each one through the two PAQ questions as honestly as he can. Knowledge and skill. Motivation and desire. He plots each one.

- Six were genuine Quadrant 2 (***Knowledge/Training***) problems. Employees who wanted to perform and lacked the specific knowledge or skill to do it. Training was the right answer for those six. He feels good about those.

- Three were Quadrant 3 (***Rewards/Recognition***) motivation problems. Employees who knew exactly how to do the job and were not doing it. Two of them were on teams where the incentive structure rewarded the wrong behavior. One was a tenured supervisor who had stopped caring after a promotion he felt he deserved went to someone else. None of them needed training. All three needed a conversation that nobody had been willing to have.

- Two were Quadrant 4 (***Resources/Environment***) problems. A pick rate issue that turned out to be a workflow sequencing problem introduced in a system upgrade six months ago. A safety compliance gap that traced back to equipment that had not been properly maintained. Training had not fixed either of them. The root cause was never the people.

- Three were Quadrant 1 (***Selection/Coaching***) problems. Employees who lacked both the capability and the motivation to perform in the roles they were in. Carlos had known this about two of them for over a year. He had requested training twice for each of them because it felt like the more supportive option. It was not. It was the more avoidant one.

He sends Priya a message before she arrives at her desk. *"I think we need to redesign how I submit training requests. Can we talk this week?"*

That message is not a small thing. It is the beginning of a different kind of partnership between operations and HR. One that starts with a question instead of a solution. One that costs fifteen minutes and saves six weeks of course development that was never going to move the needle.

Carlos did not need more training. He needed better questions.

Now he has them.

When the Write-Up Arrives Before the Standard Does

The PAQ's value extends beyond training requests. Consider the moment a manager walks into HR not to request a course, but to initiate a performance improvement plan.

The HR leader's first question should not be "what documentation do we have?" It should be the same two PAQ questions asked in a different context. Does this employee have the knowledge and skill to meet the expectation? Do they have the motivation and desire to do so?

But as Joe Dunlap puts it, before you *diagnose the employee*, **confirm the standard existed**. Did the employee know what was expected of them? Not generally. Specifically. Was the expectation documented? Was it communicated? Did the employee have access to the process, the tools, and the guidance needed to meet it?

If the answer to any of those questions is uncertain, HR has just identified a **Quadrant 4 problem** masquerading as a performance management situation. The employee may be failing against a standard that was never clearly set. Documenting that failure before confirming the expectation existed exposes the organization legally and does the employee a fundamental injustice.

This is one of the most consequential places HR can apply the PAQ framework. Not as a training router. As a check on whether a performance management action is being built on solid ground. The fifteen-minute diagnostic before a write-up is not a delay. It is the difference between a defensible process and a liability.

AI can support this too. A structured diagnostic prompt run before a performance improvement plan is initiated can surface whether the expectation was documented, whether the employee had access to what they needed, and whether the gap is genuinely attributable to the individual or to the system around them. That is HR functioning not just as a compliance function but as a performance architecture partner to the business.

> **Performance Improvement** – *AI agents can currently automate only 3% of real, complex tasks. The rest? That is still a human performance problem. And without intentional design, AI risks producing what performance improvement professionals now call "AI Workslop" – content that looks finished but delivers nothing.*
>
> ~ ISPI, Performance Improvement Journal, March 2026

Chapter 8: From Diagnosis to Development: AI in Learning, evelopment

Building the Learning Infrastructure Your Organization Actually Needs

CASE IN POINT

Diane is the Head of L&D at a global manufacturing company with over 8,000 employees. She manages a team of three and a legacy learning platform that feels more like a digital graveyard than a growth engine. Sixty percent of employees have not logged in during the last six months.

Figure 8.1 - Diane's Story

The pressure arrived last week when her CEO stood up in front of the leadership team and announced that the company would become a skills-based organization by Q3. Diane was left staring at a blank spreadsheet. She has no centralized skills data, no standardized taxonomy, and zero budget to bring in a consulting firm to map the future of her workforce.

She knows what a skills-based organization actually requires: a skills taxonomy, a skills data baseline, skills-linked learning pathways, skills-informed succession planning, and manager capability to have development conversations that go beyond the annual review. None of these things exists in her organization right now.

She also knows something that most of her peers do not. She has spent the last month working through the performance gaps in her organization with the PAQ framework and an HRBP who finally asked the right questions before accepting the next training request. She knows which problems are Quadrant 2 Knowledge/Training problems that learning can actually solve. And she knows which ones are not.

That diagnostic clarity is new. It changes everything about where she starts.

What Diane does not yet have is the infrastructure to act on that clarity at the scale her organization requires. She cannot write 8,000 personalized learning plans by hand. She cannot build a skills taxonomy from scratch in six weeks with a team of three. She cannot run a meaningful organizational network analysis or a change readiness assessment without tools that did not exist in her budget two years ago.

For Diane, AI is not a luxury or a nice-to-have innovation.

It is the infrastructure that makes the diagnosis actionable. The force multiplier that gives a team of three the capability to deliver what her organization has never had: learning that is precise, personalized, and connected to a real understanding of why the performance gap exists in the first place.

This chapter gives her the tools to build it.

***

AI in L&D – From Content Delivery to Capability Orchestration

The PAQ tells you which problem you are solving. The applications in this section tell you how AI helps you solve it. Once the diagnostic has confirmed a genuine knowledge or skill gap, Quadrant 2, these are the tools that close it.

In most HR functions, AI creates risk that must be carefully managed. In L&D, used well, it creates capability. That is the distinction that makes this chapter different from every chapter before it. Traditional L&D pushed content at scale and hoped for impact. A course was designed for a job category and rolled out to everyone in that category. Completion was measured. Impact was assumed. The model was efficient in a narrow sense. It consumed HR budget and produced learning records. What it rarely produced was changed performance.

AI allows us to move from content delivery to capability orchestration. Not learning designed for roles. Learning designed for people. The difference is not cosmetic. Most learning fails because it is designed for an average role that nobody actually occupies. Capability grows when it is designed for the specific person, in the specific gap, at the specific moment.

For Diane, the question is not whether AI can help. It can. The question is where to start, how to build on a foundation of almost nothing, and how to deliver something real against a CEO's Q3 deadline without creating data privacy risks, engagement problems, or a skills taxonomy that becomes obsolete the moment it is published.

The answers are here. And they are more accessible than Diane thinks.

Personalized Learning Pathway Generation

The highest-value application of AI in L&D is also the simplest to explain: taking what you know about an individual employee, their role, their skills gap, their learning history, and their career aspiration, and generating a personalized learning pathway that is relevant to them specifically rather than prescribed for their job category generically.

Generic learning programs fail for a predictable reason: they are designed for an average learner who does not exist. The finance manager who needs to build commercial influencing skills and the finance manager who needs to build data literacy skills are both finance managers. A generic finance learning pathway serves neither of them well. An AI-generated pathway built around their specific gap, their preferred learning format, and their time availability serves both of them significantly better.

The practical starting point for most L&D teams is the skills gap profile: a description of where the employee is now against where the role or their career aspiration requires them to be. AI takes that profile and generates a pathway with specific resources, a sequence, a timeframe, and milestones. The L&D professional reviews it, adjusts it, and makes it available to the employee and their manager. The AI does the curation. The human does the contextualization and the conversation.

Content Curation and Recommendation Engines

The learning content marketplace is one of the most overcrowded spaces in HR technology. Platforms like LinkedIn Learning, Coursera, Udemy, and Degreed sit alongside internal libraries, YouTube, podcasts, articles, and white papers. The problem is not access. It is volume. Content has grown faster than any L&D team can evaluate, curate, or recommend it intelligently.

AI is genuinely useful here as a curation engine: identifying the highest-rated, most current, most relevant resources for a specific skill or topic from across multiple sources, filtering by format preference, and ranking by quality signals. This is not a judgment task. It is a pattern-matching task. It is exactly what AI does well.

The quality review gate is clear. If L&D recommends it, a human reviews it. Not everything needs review. Only what carries the L&D endorsement. That content must be accurate, current, and aligned with the organization. AI curates. Humans endorse.

AI-Generated Learning Content

AI can generate learning content such as explainer texts, quiz questions, scenario descriptions, case studies, facilitator guides, and infographics. For L&D teams producing large volumes of content with small teams, this is a genuine time-saving application. It is also the application with the highest quality risk in L&D.

AI-generated learning content inherits the hallucination problem from Chapter 2. Content that is factually incorrect, procedurally inaccurate, or jurisdiction-specific in a way the AI does not recognize is a meaningful risk in a learning context. The risk is highest where accuracy is not a preference but a requirement: compliance training, technical skills content, and regulatory content. In those areas, AI-generated content must be reviewed by a subject matter expert before it reaches a learner. That is not optional. It is the human checkpoint that keeps this squarely in Zone 2.

The quality review process for AI-generated learning content is non-negotiable. Every piece of AI-generated content deployed to employees must be reviewed by a subject matter expert before deployment. Not a general edit. A substantive accuracy review. For compliance and technical content, this means review by someone who could have written the content themselves. The AI saves the first 70% of the effort. The human expert provides the final 30%. And that 30% is where the liability lives.

Skills Gap Analysis Using AI

Skills gap analysis has traditionally been one of the most resource-intensive L&D activities: survey design, data collection, analysis, reporting, and action planning, repeated across multiple roles and departments. For a team of three like Diane's, a rigorous skills gap analysis at an organizational scale was practically impossible without external support.

AI changes this significantly. The process for building a skills baseline without a data science team or a large budget is described in Section 7.6. Here, the headline: AI can reduce the time required to build a meaningful skills gap picture from months to weeks, and from requiring specialist analytical skills to being manageable by any L&D professional with a clear question and a structured prompt.

One PAQ note worth making explicit: a skills gap analysis that surfaces a fundamental mismatch between an employee's capability profile and the role they occupy is not a training finding. It is a Quadrant 1 selection finding. The data may point toward a placement or hiring conversation rather than a development plan. The L&D practitioner who can make that distinction and bring it to the HRBP or hiring manager with confidence is operating as a performance consultant. That is exactly the professional identity this chapter is building toward.

Agentic L&D: Where Automation Genuinely Belongs

The agentic applications of AI in L&D are some of the cleanest in the entire HR function because they are genuinely Zone 1 tasks. Learning reminder workflows. Compliance training deadline tracking and escalation. Certification expiration notifications. Course completion confirmation and badging. These are administrative processes that follow defined rules, produce reversible outcomes, and require no human judgment to execute correctly.

The governance requirement is the same as for all agentic systems from Chapter 4: map the workflows, identify the triggers, build the escalation rules for edge cases, and set the post-execution audit cadence. An employee on leave who misses a compliance deadline needs a human response, not an automated penalty trigger. The Zone 1 classification holds with appropriate monitoring.

One distinction worth drawing clearly: agentic L&D workflows manage the administrative layer of the learning function. They do not diagnose performance problems or prescribe interventions. An agentic system that automatically enrolls an employee in a training course because their performance rating dropped is not a Quadrant 2 intervention. It is an automated assumption that the performance gap is a training problem, which is exactly the assumption the PAQ exists to challenge. Automation belongs in the administrative layer. Diagnosis belongs with the human practitioner.

Building a Skills-Based Organization With AI

The skills-based organization the CEO is asking Diane to build requires two things that most organizations treat as separate workstreams but that belong together. It requires skills data as to what people can do. And it requires performance diagnostic capability as to why people are not doing it. The PAQ provides the second. This section builds the first. Neither is sufficient without the other. A skills taxonomy without a diagnostic framework is a library with no librarian. A diagnostic framework without skills data is a conversation with no evidence. Together they are the foundation of a genuine skills-based organization.

What a Skills-Based Organization Actually Means

A skills-based organization is one in which talent decisions, such as; hiring, development, deployment, succession, and reward are made on the basis of verified skills rather than job titles, tenure, or credentials. The logic is compelling: job titles are poor proxies for what people can actually do. Tenure is a poor proxy for what people currently know. Credentials are a poor proxy for current capability in a rapidly changing skills landscape.

What AI can help with in the journey to skills-based learning: building the taxonomy, generating the self-assessment, curating the learning recommendations, and modeling the skills needs of future roles. What AI cannot do: validate skills through observation, assess the quality of a skill in practice, or replace the human conversation about development and career aspiration that is the foundation of a genuine skills-based culture.

AI can map skills. It cannot judge potential. A skills-based organization is not built with a taxonomy. It is built into conversations. The taxonomy is the foundation. The manager conversations are the structure. AI gets you the foundation faster than you could build it alone. The rest is still human work.

The Skills Data Problem and How to Start Anyway

Most organizations that want to become skills-based face the same foundational challenge: they do not have the skills data they need to operate that way. Job descriptions are inconsistent. Performance ratings reflect manager bias as much as employee capability. Self-assessments are unreliable without a validated framework to anchor them. And the idea of running a comprehensive skills assessment across hundreds or thousands of employees feels overwhelming without a dedicated analytics team.

The insight that unlocks this for teams like Diane's: you do not need clean, structured data to begin. You need a starting point. An imperfect skills taxonomy built from existing job descriptions is significantly more useful than no taxonomy at all. A self-assessed skills baseline with known limitations is significantly more useful than no baseline at all. The journey to a mature skills-based organization starts with a usable first version, not a perfect one.

The Six-Step AI-Assisted Skills Taxonomy Build

The process below takes a team of one or two L&D professionals from no skills data to a validated skills taxonomy and a self-assessed skills baseline in approximately six weeks, using AI as the primary analytical engine and managers as the validation layer.

Job descriptions are the practical starting point for most organizations because they are immediately available. Where workflow documentation exists, it is a stronger input. A job description captures what a role was designed to require. A documented workflow captures what the work actually demands under real conditions. Those two things are not always the same. Where both exist, use both. Where workflow documentation is absent, the taxonomy build is also an opportunity to surface that gap and begin closing it.

Step	Action	How	Output
Step 1	**Extract skills from job descriptions**	Paste your 10–20 most representative JDs into an AI tool. Ask it to identify the skills, competencies, and knowledge areas each role requires. Deduplicate and cluster the output.	Raw skills inventory
Step 2	**Cross-reference with industry frameworks**	Compare your extracted skills against published frameworks (SFIA, O*NET, sector-specific standards). Ask AI to identify gaps and emerging skills not yet in your JDs.	Validated skills framework
Step 3	**Build the taxonomy structure.**	Organize skills into three tiers: foundational (every employee), role-specific (by function), and emerging (strategic priority for next 3 years). AI drafts, L&D team reviews and edits.	Draft skills taxonomy
Step 4	**Validate with managers**	Share the draft taxonomy with 5–10 business line managers. Ask: "Does this capture what excellent performance in your team actually requires?" Incorporate their edits.	Manager-validated taxonomy
Step 5	**Deploy the self-assessment**	Use AI to convert the taxonomy into a self-assessment tool. Employees rate themselves against each skill. AI generates personalized gap reports and learning recommendations.	Skills baseline + learning plans
Step 6	**Set the monitoring cadence.**	Repeat the self-assessment annually, or after major role or technology changes. Build manager validation into the performance review cycle.	Living skills data

Figure 8.2 - Six-Step AI-Assisted Skills Taxonomy

Skills-Based Succession Planning

Succession planning has traditionally been conducted through a combination of manager nomination and leadership team discussion, a process that is heavily influenced by visibility, relationships, and the cognitive biases of the people in the room. AI-assisted succession planning does not replace that process. It enriches it by adding a skills dimension that subjective nomination alone cannot provide.

The AI-assisted succession planning framework works as follows. For each critical role, define the skills profile required. Map current employees' validated skills profiles against that requirement. AI identifies the gap for each potential successor and generates a targeted development plan to close it. The succession decision, meaning who is nominated, who is supported, and what the timeline is, remains entirely human. AI provides the skills evidence. Humans make the judgments.

One grounding note on succession readiness. Skills profiles and development plans describe what someone is capable of becoming. Succession readiness is ultimately demonstrated in execution, not inferred from capability models alone. The most reliable evidence of readiness is observed performance under real conditions, in the work itself, not in a validated competency framework. AI can identify the gap and generate the development plan. Only the work can confirm the readiness.

A PAQ lens applies here too. A successor candidate whose skills gap is large but whose motivation and desire to grow into the role is high sits in Quadrant 2. Development investment is warranted. A successor candidate whose skills gap is modest but whose motivation to take on the role is genuinely low sits in Quadrant 3. More development content will not close that gap. A conversation about aspiration, consequence, and fit will. Succession planning that ignores the motivation dimension produces development plans that look rigorous on paper and produce no movement in practice.

THE HUMAN JUDGMENT LAYER IN SKILLS ASSESSMENT

These skills require human observation, not algorithmic inference:

Leadership presence and executive gravitas >>> observable in the room, not in the data
Interpersonal judgment under pressure >>> requires direct observation across contexts
Values alignment and ethical decision-making >>> revealed through behavior over time
Collaboration quality >>> how someone works with others, not how often they interact
Communication effectiveness >>> the impact on the audience, not the frequency of communication
Creative problem-solving >>> the quality of the thinking, not the volume of ideas generated

AI can tell you that someone completes their training on time.
It cannot tell you whether they are ready to lead.
That judgment requires a human who has watched them work.

Figure 8.3 - Skills Requiring Human Judgement

AI in Organizational Design and Development

This is where the macro and the micro finally meet.

The sections before this one have built the case for AI-enabled performance consulting at the individual and team level. The PAQ diagnoses the Job/Performer gap. The skills taxonomy builds the data foundation. The human performance consultant asks the questions the organization has never had the infrastructure to ask consistently.

Organizational Development has always worked the other side of that equation. Culture, structure, strategy, network analysis, and change readiness are organizational and process level work.

It is important work. But for decades it has operated largely disconnected from the individual and team level diagnostic work the PAQ represents. OD could tell you the organization had a performance culture problem. It could not tell you why this person, on this team, in this department, was not performing.

AI connects these levels for the first time at operational scale. The micro-level PAQ diagnostics that L&D practitioners are now running feed upward into the organizational performance picture that OD needs to work with. The macro-level OD findings translate downward into individual intervention recommendations that practitioners can act on. The three *Rummler-Brache* levels, Organization, Process, and Job/Performer, finally talk to each other in real time.

This is what *Peter Senge's* Learning Organization always required and never had. Not more training. Not a better LMS. A diagnostic and feedback infrastructure that operates at every level simultaneously. AI is that infrastructure. And the OD applications below are where it becomes most visible at the organizational level.

Organizational Network Analysis

Every organization has two structures: the formal one, visible in the org chart, and the informal one, invisible in the org chart but responsible for most of how work actually gets done. Organizational network analysis maps the informal structure, the people others go to for advice, expertise, information, and support, and surfaces the influencers, connectors, and potential bottlenecks that no org chart reveals.

AI-assisted organizational network analysis typically works by analyzing communication metadata, including email headers, calendar data, and collaboration platform activity, to map interaction patterns across the organization. It does not read the content of communications, or should not, which is a consent and privacy question addressed in the Grey HR Zone framework at the end of this chapter. It analyzes the patterns: who communicates with whom, how frequently, and in what direction.

The output is a network map that OD professionals use to answer questions like: who are the informal influencers in this part of the organization? Where are the silos? Who is overloaded as an informal knowledge hub? Who is isolated in ways that their performance data does not yet reflect? These are valuable OD insights. The governance requirement is straightforward: employees must know that their communication patterns are being analyzed, what data is used, and what decisions the analysis informs.

One interpretive caution worth building into your ONA practice. Communication patterns do not always reflect influence or collaboration preferences. They often reflect workflow design. An employee who appears as a bottleneck in the network map may be a bottleneck because the workflow routes every decision through their role, not because of anything about them personally. Before drawing conclusions about individuals or teams from ONA data, ask whether the pattern reflects people or process. That distinction changes both the diagnosis and the intervention significantly.

Change Readiness Assessment

Before a major transformation such as a merger, a restructuring, a technology change, or a strategy shift, understanding the organization's emotional and cultural readiness for change is one of the most valuable inputs an OD function can provide. AI-assisted sentiment analysis, applied to pulse survey data, engagement tool outputs, and anonymized communication signals, can give OD professionals a richer and more current picture of change readiness than a periodic engagement survey alone.

The honest framing is this: AI sentiment analysis in this context is a signal aggregator, not a truth-teller. It tells you that anxiety signals are elevated in a particular department, or that language patterns in a specific team suggest resistance. It does not tell you why or what to do about it. The diagnosis requires human interpretation. The intervention requires human presence. AI improves the quality of the evidence. OD professionals make the clinical judgment.

Culture Measurement With AI

Culture is the pattern of assumptions, beliefs, and behaviors that shapes how an organization actually operates, as distinct from how it says it operates. AI can measure certain aspects of expressed culture: the language used in internal communications, the themes that emerge in open survey responses, and the patterns of behavior that show up in collaboration data. What it cannot do is measure the things that matter most about culture, whether people feel psychologically safe, whether stated values are lived in practice, whether leadership behavior reinforces or contradicts the culture the organization claims to have.

AI is the thermometer, not the weather. It tells you what temperature the room is. It cannot change the climate. Culture measurement with AI gives OD professionals better evidence to bring into the human conversations that culture change actually requires. The diagnosis is richer. The intervention is still human.

Workforce Planning With AI

AI workforce planning tools do their best work in scenario modeling. Given the current workforce composition, attrition trends, skills development rates, and hiring capacity, what does the workforce look like in two or three years under different strategic assumptions? This is genuinely valuable for organizations facing significant change, whether digital transformation, demographic shift, market entry, or the kind of reskilling challenge that Diane's CEO has just named as a strategic priority.

The data quality dependency from Chapter 2 applies here with particular force. Workforce planning AI is only as good as the data it is modeling. Inconsistent job titles, patchy attrition data, and skills data that reflects what people were hired to do rather than what they can currently do will produce workforce plans that are precise but not accurate. Building the skills taxonomy in the previous section is not just an L&D exercise. It is the data foundation that makes AI workforce planning meaningful.

The Limits of AI in OD

The OD applications of AI are diagnostic, not generative. AI can tell you something is wrong, or different, or changing. It cannot tell you what to do about it. And there are dimensions of organizational health that AI cannot measure at all. Not because the data is missing, but because the thing being measured does not exist in a form that data can capture.

Organizational identity, what an organization fundamentally is and stands for, cannot be algorithmically designed. Values cannot be inferred from behavior patterns and then fed back as recommendations. Culture cannot be optimized by an AI system any more than an algorithm can optimize a community. These are human endeavors. They require human leadership, human conversation, and human judgment about what matters.

The organizations that use AI well in OD use it to generate better evidence for human judgment, not to replace it. The question is never what does the AI say we should do. It is what does the AI's evidence tell us that we should be paying attention to.

And when OD asks that question at the organizational level while L&D practitioners are asking the PAQ questions at the individual and team level, the Learning Organization Senge described in 1990 is no longer an aspiration. It is an operating model.

Cross-Industry Case Study: Mid-Size U.S. Healthcare System

CROSS-INDUSTRY CASE STUDY

Organization: Mid-size U.S.-based healthcare system (~1,200 employees)
Function: Learning & Development (No dedicated OD or People Analytics function)

The Problem:

The organization identified that **~40% of its workforce (~approximately 480 employees)** across clinical support, patient services, and administrative functions would require **significant reskilling for digital and data-enabled roles within three years.**

The CHRO commissioned a **skills baseline and reskilling roadmap.**

There was just one problem.

- No skills taxonomy
- No centralized skills data
- No budget for external consulting or assessment tools
- No data science capability
- No dedicated OD or workforce planning function

The responsibility landed where it often does in mid-size healthcare organizations: with **Learning & Development.**

The L&D team, with two people responsible for training, onboarding, compliance education, and anything resembling workforce capability, understood what was required: a skills taxonomy, a baseline, and a way to connect learning to real capability gaps.

They also understood what they did ***not*** have: time, budget, or the option to wait.

A traditional consulting-led approach was estimated at **$350,000–$500,000 and 9–12 months.**

The organization was actively rolling out new systems and workflows. Waiting was not a strategy.

What They Did:

The L&D team used an AI tool to extract skills and competencies from existing job descriptions across **30+ role families,** including clinical support, patient services, and administrative roles.

They cross-referenced the output against **SFIA (Skills Framework for the Information Age)** and internal role expectations to build a **draft skills taxonomy (~180 skills)** across three tiers:

- Foundational
- Role-specific
- Emerging digital

Week 3: Validated the draft taxonomy with 8 senior managers across the highest-priority functions and incorporated their edits. 23 skills were modified; 11 were added; 9 were removed as not relevant to actual role performance. This step grounded the model in **how work actually gets done in a healthcare environment.**

Weeks 4–6: Converted the validated taxonomy into a self-assessment tool. Deployed to employees with a covering communication from the CHRO explaining the purpose, the data use, and the employees' control over their own data. AI-generated personalized learning recommendations for every employee based on their self-assessment results.

Total cost: 3 weeks of L&D team time and approximately **$200** in AI tool usage.

Outcome:

- Skills baseline established for 1,200 employees in 6 weeks
- 340 employees in the highest-priority reskilling cohort had personalized learning plans in place within the first month
- L&D platform engagement increased by 48% in the first quarter following deployment, which was attributed to the relevance of personalized recommendations compared to the previous generic catalog approach
- The CHRO presented the skills baseline to the board at the quarterly meeting, seven months ahead of the original consulting engagement timeline

The Lesson:

The skills data problem is not solved by waiting for a perfect data collection process. It is solved by starting with what you have: job descriptions, industry frameworks, and manager knowledge, and using AI to build a useful first version fast. An imperfect skills taxonomy that exists is worth more than a perfect one that is still being scoped.

GREY AREA DECISION FRAMEWORK

Grey Area: AI-Inferred Skills and Performance

This is the grey area that is most specific to L&D and OD. And the one that most easily goes unnoticed until it becomes a trust or legal issue.

The scenario: AI tools that infer skills, potential, or performance readiness from indirect signals such as email response time, meeting participation frequency, Slack activity, calendar patterns, and document collaboration data, without employees' knowledge or consent.

Why this is contested:

- **The inference is unreliable:** communication frequency is not a proxy for communication quality. Meeting attendance is not a proxy for contribution. Email response time is not a proxy for engagement.
- **The inference is invisible:** employees do not know their inferred score exists, cannot challenge it, and may be disadvantaged by it in development or succession decisions without any opportunity to provide context.
- **The inference may be discriminatory**: communication patterns vary significantly by role, language background, seniority, and caregiving responsibilities.

An AI that infers potential from communication volume will systematically disadvantage employees whose communication patterns differ from the norm for reasons unrelated to their capability.

The transparency test:

"Would your employees be comfortable knowing exactly what data the AI is using to make inferences about them, and what decisions those inferences influence?"

If the answer is no, or even uncertain, do not use it.

What is acceptable:

- Skills self-assessment with AI-generated recommendations, employee-initiated and employee-visible.
- Learning completion data as a skills indicator, transparent and employee accessible.
- Manager-validated skills profiles, meaning human judgment, applied to an AI-generated starting point.
- Aggregated, anonymized organizational network analysis data used at the team or function level, not the individual level.

What is not acceptable:

- Using AI to infer individual performance or potential from communication patterns without explicit employee awareness
- Feeding behavioral inference data into succession or promotion decisions without disclosure
- Storing inferred skills profiles in employee records without consent

The consent rule:

- Any AI tool that analyses individual employee behavior for development or performance purposes requires explicit employee awareness of what is being analyzed, why, and how the output will be used.
- In the EU, UK, and many other jurisdictions, this also requires a legal basis under data protection law. interests are not automatically sufficient for processing sensitive behavioral data. Involve your DPO before deployment.

CHAPTER CHECKLIST

☐ I have identified the top three L&D workflows where AI could save significant time this quarter and mapped each one to Zone 1, 2, or 3.

☐ I have a quality review process for any AI-generated learning content before it reaches employees, including a named subject matter expert for compliance and technical content.

☐ I have begun building a skills taxonomy, even a draft version, using the **six-step process** in this chapter.

☐ I have checked that any AI tool used in L&D for individual employee data complies with our **data privacy obligations**, and **our employees know it is in use**.

☐ I have a plan for the human judgment layer in our skills assessment process by identifying which skills require manager observation and which can be self-assessed.

☐ I have identified the top three performance problems in my organization that were addressed with training in the last quarter and run each through the PAQ to confirm training was the right intervention. Where it was not, I have documented the correct quadrant and the intervention that should have been recommended instead.

☐ I have assessed whether my L&D practitioners have the diagnostic tools, the AI-assisted capability, and the organizational confidence to function as human performance consultants, not just content producers. If not, I have identified the first step toward closing that gap.

PROMPT LIBRARY & WORKBOOK EXERCISES

SKILLS TAXONOMY PROMPT

[Build a skills taxonomy for a [*role title*] in a [*industry*] organization. Include: core technical skills (what they need to know and be able to do technically), behavioral competencies (how they need to work with others and navigate complexity), and emerging skills likely to be needed in the next 3 years based on industry trends. Format as a structured framework with three tiers: foundational, role-specific, and emerging. Include 3–5 behavioral indicators for each competency.]

LEARNING PATHWAY PROMPT

[Based on this skills gap profile: [*describe the gap, such as; role, current capability, target capability*], generate a personalized learning pathway. Include: recommended resources by type (online course, article, video, on-the-job experience, coaching or mentoring), estimated time investment for each, and 30/60/90-day milestones. Prioritize resources that are free or low-cost. Flag any resources that require manager facilitation.]

WORKBOOK EXERCISE

List the **top five capabilities** your organization will need most in three years. For each capability, rate your organization's current collective capability from 1 to 10. Multiply the importance of each capability by the inverse of your current capability score (10 minus your rating). The highest scores are your priority L&D investment areas and your AI pilot starting points.

TEAM EXERCISE: THE ONA LITE

Run a manual organizational network analysis with one team. Ask 10–15 people: '*Who do you go to most often for advice or information when you have a challenging problem*?' and '*Who comes to you*?' Map the results. Identify the informal influencers, the people who appear most often as a resource for others, regardless of their formal seniority. These are your **change champions for AI adoption** in this team, and the people whose buy-in is most valuable for any L&OD initiative.

SKEPTIC'S CORNER FAQ

Q: "Our learning platform already has AI recommendations built in. Do we need to do anything else?"

A: Yes. Built-in AI recommendation engines are AI tools. They need the same scrutiny as any other AI tool. Run a demographic disaggregation check on who is being recommended, which learning, and at what frequency. Skewed recommendations are common when recommendation engines are trained on historical engagement data. They compound existing development inequity by directing more learning resources to employees who already have higher engagement rates. The bias audit process from Chapter 5 applies here.

Q: "We don't have skills data. Where do we even start?"

A: Start with your job descriptions. They are an imperfect but immediately available proxy for the skills your organization values. The six-step taxonomy built in this chapters starts there and builds outward, adding manager validation and employee self-assessment to create a progressively more accurate picture. You do not need clean, structured data to begin. You need a question, a set of job descriptions, and an AI tool. That is enough to produce a useful first version.

Q: "Can AI really help with culture and OD? Isn't that inherently human?"

A: The OD applications of AI are diagnostic, not generative. AI can measure indicators and aggregate patterns. It cannot define your values, design your organization's identity, or sustain your culture. Think of AI in OD as a better measurement instrument. It tells you where to look, what questions to ask, and where the human conversations most urgently need to happen. Culture is still and will always be human work. AI helps you find the right room to have the conversation in.

What Diane Builds by Q3

Diane does not wait for a consulting budget that will not arrive in time. She starts with what she has: job descriptions, three weeks of focused effort, and an AI tool that costs less per month than a team lunch.

By the end of week two she has a draft skills taxonomy. Not perfect. Not complete. But validated by five plant managers who looked at it and said yes, this is what good looks like in our environment. That has never existed before in this organization. It exists now because Diane stopped waiting for the right conditions and started building with the conditions she had.

By the end of week four she has a self-assessed skills baseline for the priority population her CEO identified. The AI generated the assessment from the taxonomy. Employees completed it in twelve minutes on average. The gaps are visible for the first time. Not assumed. Not anecdotal. Visible.

By the end of week six she has personalized learning plans for every employee in the priority group. The platform engagement numbers are moving for the first time in a year. Not because the content improved. Because the content is now connected to a real diagnosis of a real gap in a real person's performance. The recommendations are relevant. Employees notice when something is relevant to them specifically.

She presents the CHRO with data she did not have two months ago. Not perfect data. She is clear about its limitations. But real data, grounded in manager input, built on a validated framework, and already producing learning activity that would not have happened without it.

The CEO gets his skills-based organization. Not the finished version. The foundation. And a credible roadmap to the finished version that is grounded in evidence rather than aspiration.

And Diane gets something she did not expect.

Operations calls her before they submit the next training request. Not because she asked them to. Because Priya shared the PAQ framework with the plant managers after the work they did together in Chapter 7. And one of those managers ran it on his own before he picked up the phone. He called Diane not to order a course. He called to ask what the right intervention was for a Quadrant 3 problem on his floor.

That call is not a small thing. It is the beginning of something the function has always said it wanted to be but has rarely had the infrastructure to become.

Diane's team of three did not become five. It became different. AI handles the content. The practitioners handle the diagnosis and the relationship. And the organization finally gets what it has been paying for and rarely receiving: performance change that lasts.

This is not a technology story. It is a capability story. And the capability that matters most is not the AI.

It is the human who finally has the tools to use it well.

Worth Remembering

Organizations are spending more. Training fewer hours. And proving almost none of it works.

> **Training Dollars Spend** – *U.S. organizations spent $102.8 billion on training in 2025. The average employee received 40 hours, down from 47 hours the year before. And only 4% of organizations excel at using learning data to inform business decisions. More investment. Less time. Almost no proof.*
>
> ~ Training Magazine, 2025 Training Industry Report; ATD, The Future of Evaluating Learning and Measuring Impact, 2025

Chapter 9: Change Management - Leading AI Adoption Without urning Out Your Team

CASE IN POINT

Keisha is a People Director at a logistics company with 2,200 employees. She has been in HR for sixteen years. She is not a technophobe. She championed the AI performance management tool she is now watching fail in slow motion.

Figure 9.1 - Keisha's Story

Six months ago she presented the business case. She walked the executive team through the vendor demo. She managed the procurement process. She designed the rollout communications. She genuinely believed in the tool, and she still does.

Her HR team hate it. They say it produces generic outputs that strip the nuance from their professional judgments. They are spending more time editing AI drafts than they would have spent writing from scratch. Her managers distrust it. They say the performance ratings feel algorithmic and impersonal. Two senior managers have stopped using the manager-facing dashboard entirely. Her employees are convinced it is being used to build a case for redundancies. Nobody told them that. Nobody denied it either, because nobody thought to say anything until the rumors were already running.

Keisha believes in the tool. But she is losing the room. And the hardest part is that she knows the technology is not the problem. The people process is.

This chapter is about winning the room back. And doing it very differently next time.

***

Every chapter in this book has been, in some dimension, about governance. About what humans must protect, oversee, and decide. This chapter is about something that is both simpler and harder: the human beings who have to live inside the AI adoption you are leading.

The biggest barrier to successful AI adoption in HR is not technology. It is people. And the people most at risk of becoming that barrier are not your employees. They are your HR team.

You are not leading an AI rollout. You are leading a trust negotiation.

That reframe matters. Because trust negotiations require a fundamentally different set of skills than system implementations. They require honesty about uncertainty. They require genuine listening before communication. They require the willingness to acknowledge when something is not working before the evidence becomes impossible to ignore. They require credibility, which is the one thing an AI rollout can destroy faster than any other change initiative.

The biggest mistake organizations make in AI adoption is treating it as a communication problem. It is not. It is a credibility problem. The organizations whose AI adoption stalls are almost never the ones that communicated too little. They are the ones that communicated reassurance they could not substantiate, enthusiasm that did not match lived experience, and certainty about outcomes that were never certain.

Change management in the age of AI lives almost entirely in the Grey HR Zone. Not because the steps are unclear, but because the emotions are. The fear is real, even when it is not rational. The resistance is meaningful, even when it is inconvenient. And the HR leader who names that honestly, who leads through the grey rather than around it, is the one whose team actually crosses to the other side.

This chapter is not a standard change management toolkit with AI vocabulary added. It is a framework built specifically for the psychological and organizational dynamics of AI adoption: the fear, the anxiety, the competence threat, and the trust deficit that make AI change different from every other change initiative your organization has managed.

Why AI Change Management Is Different

Change management has been practiced and written about for decades. The frameworks are well-established. Most HR leaders know them: build a coalition, communicate early and often, identify your champions, address resistance, measure adoption. These frameworks work for most change initiatives. They work less well for AI adoption. Here is why.

The Unique Psychological Dynamics of AI Adoption

Most change initiatives ask people to do something differently. AI adoption asks people to reckon with the possibility that what they do might no longer be necessary. That is a categorically different psychological experience, and it requires a categorically different change management response.

Three psychological dynamics are specific to AI adoption and routinely underestimated in change management planning:

- **Job threat perception**. The fear that AI will eventually replace the role, the function, or the profession. This fear does not need to be rational to be real. It does not need to be supported by evidence to shape behavior. And it does not respond to reassurance that is not grounded in specific, honest information about what will and will not change.

- **Competence anxiety**. The fear of being exposed as less capable than a machine. This is particularly acute for experienced professionals who have built their identity and their organizational value on expertise that AI can now partially replicate. A manager who has written performance reviews for twenty years and is now told to use AI assistance may experience this not as support but as an implication that their judgment is no longer trusted.
- **Trust deficit**. The fear that this is just another overpromised technology initiative that will disrupt work without delivering real value. Most organizations have a history of implementations that promised more than they delivered. When employees encounter AI hype, they filter it through that experience. Skepticism is not obstruction. It is accumulated institutional memory, and it requires evidence, not enthusiasm, to overcome.

The Prompt Burnout Phenomenon

Chapter 3 introduced prompt burnout in the context of small HR teams. It deserves fuller treatment here because it is one of the most significant hidden costs of poorly managed AI adoption, and one of the most common reasons rollouts like Keisha's stall.

Prompt burnout occurs when AI adoption is mandated rather than chosen, and when the tools deployed are not yet good enough to deliver on the efficiency promise. The person using the tool spends more time prompting, reviewing, correcting, and re-prompting than they would have spent completing the task without AI assistance. The net result is negative productivity, increasing frustration, and a team that has reached a rational conclusion: this tool is making my work harder, not easier.

The solution to prompt burnout is not better prompting training. It is better tool selection, better use case matching, and most importantly, the honesty to acknowledge when a tool is not delivering what was promised. A People Director who has championed a tool has a structural incentive to defend it even when the evidence suggests it is not working. Keisha's situation is partly a product of that dynamic. Recognizing it is the first step to addressing it.

Why the Standard Change Management Playbook Underserves AI Transitions

Standard change management frameworks were designed for transitions with clear before-and-after states: we used to do X this way, now we do it this way. AI adoption is a continuous, uncertain transition. The before state is relatively clear. The after state is genuinely unknown, because AI capabilities are evolving, organizational use cases are emerging, and the implications for specific roles and functions are not yet determinable with confidence.

Communicating a change when you do not yet know what the change will ultimately look like requires a different approach than communicating a migration to a new system. The honest answer to "What will this mean for my role in two years?" is often "We do not fully know yet." Standard change management playbooks are not designed for that answer. They are designed for transition certainty. AI adoption rarely provides it.

The HR Leader's Dual Role

HR leaders occupy an unusual position in AI adoption: they are simultaneously the change agents leading the transition and the change subjects experiencing it. They are expected to communicate with authority and confidence about a technology that many of them are still learning. They are expected to address employee fears about job security while managing their own. They are expected to model AI adoption while privately questioning whether the tool is as good as the vendor promised.

Naming this dual role is not a vulnerability. It is a leadership credential. Admitting that you are learning these tools at the same time as you are governing them is not a weakness. It is the first step in building a Human Firewall within your culture: a team of people who understand the technology well enough to protect your organization from its failures, precisely because they are willing to be honest about what they do not yet know.

The HR leader who says "*I am also figuring this out, and I am committed to doing it in a way that is honest and human*" is more credible, not less, than the one who performs certainty they do not feel. Chapter 9 is written for that leader.

The Adoption Curve for AI in HR

Not everyone in your organization is at the same place on the AI adoption journey. Not everyone is in the same emotional relationship with the technology. And applying the same change management approach to people in fundamentally different psychological positions will produce, at best, mixed results.

The five stakeholder archetypes below are not personality types. They are positions on the adoption curve, positions that people can and do move through as circumstances, evidence, and experience change. Your job is not to sort people into permanent categories. It is to understand where someone is right now and meet them there.

Archetype	How They Present	Engagement Strategy	The Key Insight
The Enthusiast	Excited, early adopter, already experimenting	Channel their energy toward high-value use cases. Involve them in governance design. Use them as peer champions but manage the expectation gap before it creates credibility problems.	Watch for Zone 3 creep: enthusiasts are the most likely to move Zone 3 decisions into Zone 1 without realizing it. Their energy needs a governance channel.
The Sceptic	Unconvinced but not hostile. Has genuine questions.	Validate their concerns explicitly. Involve them in the vendor evaluation and governance process. Their rigor will strengthen your implementation.	Their skepticism is your quality assurance. The Skeptic who asks the hard questions in week two saves you from the crisis in month six.
The Fearful	Believes AI is coming for their job. Disengaged or anxious.	Address the job threat directly and honestly. Do not deflect with reassurance. Use the Zone Sorter together to show them specifically where their judgment is still the final word.	Silence amplifies fear. Name the concern before they do. Zone 3 is the answer to their deepest worry.
The Passive	Neither resistant nor engaged. Waiting to see what happens.	Lower the barrier to entry. Give them a small, low-stakes win early. Celebrate it visibly.	Passives move in herds. When enough peers adopt, they follow. Your job is to create the visible early wins that tip the herd.
The Saboteur	Actively undermining the rollout. Vocal opposition or subtle obstruction.	Investigate the root cause before treating it as a performance issue. Try the Central Decision Test: if the tool made the final decision about their career, would they expect a human to be involved? Their answer usually reveals the root.	Sabotage is almost always a signal about something the rollout got wrong. Find it before escalating.

Figure 9.2 - Stakeholder Archetypes

Mapping Your Stakeholders

Before designing your change management approach, map the key stakeholders in your AI adoption initiative against these five archetypes. Be honest. The temptation is to overcount Enthusiasts and undercount Fearful and Skeptic archetypes, because the latter two require more investment. Underestimating the distribution in those groups is one of the most common reasons rollouts like Keisha's encounter resistance that was predictable and preventable.

THE ADOPTION MAPPING EXERCISE

For your current AI adoption initiative, list every key stakeholder group:

- *Your HR team (individually, not as a group — they will not all be in the same position)*
- *Your senior managers / HR business partners*
- *The employee population most immediately affected*
- *Your executive sponsors*
- *Your IT and operations partners*

For each, ask: based on what I have observed so far, which archetype best describes their current position?

Then ask the harder question:
"What would I need to observe before concluding that someone is a Saboteur rather than a Fearful or Skeptic who has not yet been given the right information?"

Most apparent Saboteurs are Fearful or Sceptic stakeholders who have not been heard. The mapping exercise does not sort people. It focuses your energy.

Figure 9.3 - Adoption Mapping

Communication Scripts and Templates

The communications that determine whether an AI adoption succeeds or fails are not the launch announcement. They are the conversations that happen in the three weeks after the launch announcement, when questions emerge that the announcement did not address, when the tool does not work exactly as demonstrated, and when the rumor mill fills the gaps that official communications left.

This section gives you the communication frameworks for the moments that matter most.

The All-Staff AI Communication Framework

The all-staff communication about a new AI tool has one job: it must address what employees are actually worried about before they have to ask. That means the job security question must be in the communication, addressed directly, before it is the first question raised in the Q&A. The moment an employee has to ask whether their job is at risk is the moment you have already lost a significant part of the trust conversation.

THE ALL-STAFF COMMUNICATION STRUCTURE

1. What we are introducing and why (2 sentences AND be specific, not corporate)

Example: "We are introducing [tool name] to help our HR team spend less time on [specific Zone 1 administrative task] and more time on the conversations and decisions that our people need them for."

2. What it does and does not do (specific, not vague)

Name the Zone 1 tasks it assists with. Name the Zone 3 decisions that remain entirely human. Use that language explicitly — it connects this communication to a framework your people will recognize across every HR process.

3. The job security statement (direct, not deflecting)

This statement must be made, not implied. Example: "This tool is designed to free our HR team from Zone 1 administrative tasks so they can do more of the Zone 3 work that actually requires human judgment. We are not introducing this tool as part of any headcount reduction plan."

If a headcount reduction is planned, do not include the second sentence. Communicate honestly about what is happening

4. The human oversight statement

"No final decision regarding your pay, performance, or employment will ever be made by an AI system. A human will always make the call. Every output this tool assists with is reviewed by a named person before it affects you. [Name the specific reviewer for each workflow]."

5. The invitation to engage

"We expect questions. Here is how to ask them: [specific channel, specific person, specific timeline for response]."

What not to include: enthusiasm. An all-staff communication about AI that is full of excitement signals that the author is more interested in the technology than in the people reading it.

Figure 9.4 - All-Staff AI Communication Framework

The Manager Briefing

Line managers will receive questions from employees that HR cannot be everywhere to answer. They need briefing materials that are honest, specific, and give them the language for the questions they will actually be asked. The most common questions are predictable:

- "*Is this going to replace HR?*" Managers need a specific, honest answer, not a deflection.

- "*Why is HR using AI to evaluate my performance?*" Managers need to understand exactly what the tool does and what humans decide.
- "*Will my data be used to train the AI?*" Managers need a factually accurate answer about data use, not a reassurance that may not be accurate.
- "*I do not think the AI assessment of my team member is right. What can I do?*" Managers need to know that human override is not just permitted but expected.

The manager briefing should give a specific, plain-English answer to each of these questions. It should be delivered in a conversation, not sent as a document. And it should explicitly tell managers: "If you get a question you cannot answer, the answer is: I do not know, let me find out. Do not speculate." Speculation in a trust deficit creates more damage than acknowledged uncertainty.

The "This Is Not Replacing You" Conversation

This is the conversation Keisha did not have early enough. It is also the conversation that most HR leaders find the most uncomfortable, because it requires them to address a fear that might be partially legitimate. The honest version of this conversation depends entirely on what is actually true.

If the tool genuinely is not part of a headcount reduction plan, say so specifically: "This tool is being introduced to reduce the administrative load on the HR team. It is not part of any plan to reduce the size of the function. I am telling you this directly because I know the question is in the room, and I want to answer it before it becomes a rumor."

If the long-term implications for headcount are genuinely unknown, say that instead: "We are introducing this tool to improve how we work. I cannot tell you with certainty what the HR function will look like in three years. Nobody can. What I can tell you is that our intention is to use the capacity this creates for higher-value work, not to reduce the team. If that changes, I will tell you."

The credibility of this conversation depends entirely on its honesty. An implausible reassurance is worse than no reassurance. ***If you cannot say it and mean it, do not say it.***

Responding to a Team Member Who Refuses to Use an AI Tool

Tool refusal is more common than most AI adoption plans anticipate, and it is almost never simply obstinance. Before treating a refusal as a performance issue, treat it as a communication. The first question is: what is the specific concern?

- If the concern is competence anxiety: the response is support and low-stakes practice, not pressure. Schedule time to work alongside the person, not to train them but to use the tool together.

- If the concern is job threat: the conversation from the previous section is needed, personalized to their specific role, and honest about what you know and do not know.

- If the concern is ethical or values-based: this deserves genuine engagement, not dismissal. An HR professional who has a substantive objection to using AI in a specific workflow may be identifying a governance gap that your adoption plan missed.

- If the concern is a disability or accessibility issue: reasonable adjustment obligations apply. Is there a way to achieve the same outcome without this specific tool? If yes, that is the path. Document the accommodation and the reasoning.

The escalation path to a formal performance conversation should only be triggered after these conversations have been had, documented, and the person has been given a genuine opportunity to engage with a process that addresses their specific concern. Skipping to escalation treats communication failure as a conduct issue. That is rarely the right diagnosis.

The Executive Update

Reporting AI adoption progress to the board or executive team requires translating what is happening in human terms into the language that executive teams respond to: risk, cost, and strategic positioning.

THE EXECUTIVE AI ADOPTION UPDATE FRAMEWORK

1. **Adoption rate** (and what it tells you)

   Not just the percentage using the tool, but what the distribution across archetypes tells you about where the adoption program needs investment. Reference your current maturity stage: 'We are moving from Aware to Intentional — here is what that transition requires.'

2. **Quality of output** (not just volume of use)

   Are the outputs the tool produces being used, modified, or discarded? Usage without quality is a vanity metric.

3. **Time saving realized vs. projected**

   The honest gap between the vendor's efficiency promise and the actual time impact on your team.

4. **Governance status**

   Which workflows have human review gates in place. Which are operating without them. What you are doing about the latter. Reference the Agentic Workflow Registry if relevant.

5. **Risk register update**

   Any bias findings, compliance concerns, or employee relations issues that have emerged from AI tool use.

6. **Next quarter priorities**

   What are you doing differently based on what you have learned? Not just what is going well.

What not to include: enthusiasm metrics. The number of prompts submitted is not a meaningful adoption indicator. The quality of outcomes is.

Figure 9.5 - Executive AI Adoption Update Framework

Building Psychological Safety for AI Experimentation

The single most reliable predictor of sustainable AI adoption in HR teams is not tool quality, not training investment, and not executive mandate. It is whether people feel safe to try the tool, produce an imperfect output, and continue without penalty.

Fear of looking incompetent is the hidden brake on AI adoption in most organizations. It is the reason experienced professionals avoid the tool rather than produce results that reveal their uncertainty. It is the reason people tell you they are using the tool when they mean they opened it once and closed it. It is the reason rollouts plateau at the early adopter population and never reach the majority.

Why Fear of Incompetence Stalls AI Adoption

AI adoption is a skill acquisition process. Like all skill acquisition, it involves a period of visible incompetence before competence is achieved. A senior HR professional who has been excellent at their job for fifteen years is not accustomed to producing work that is worse than what they could produce unaided. The early stages of AI adoption frequently create exactly that experience.

In a culture where professional identity is closely tied to expertise, and where appearing to struggle with a new tool is experienced as a threat to that identity, the rational response is avoidance. The professional does not fail publicly at the new skill. They continue using their established methods, report compliance with the adoption mandate, and wait for the program to quietly move on.

Psychological safety for AI experimentation means creating a culture where the early, imperfect stage of AI use is expected, visible, and safe to admit. Where team leaders share their own clumsy prompts and their surprising failures alongside their wins. Where the conversation about AI is honest about what the tools are not yet good at, as well as what they are.

The AI Experiment Framework

The case study in this chapter demonstrates the most effective structural intervention for building adoption without mandate: the voluntary AI experiment program. The principle is simple and the evidence for it is consistent: autonomy over the experimentation process is the decisive factor in adoption, not the quality of the tool.

THE AI EXPERIMENT FRAMEWORK

Structure:

- *One hour per week of protected time for each team member to experiment with AI on a task of their own choosing*
- *The task must be real work, not a training exercise*
- *The output does not need to be used - the experiment is the point*

The weekly debrief (10 minutes, no more):

- *What did you try?*
- *What worked better than expected?*
- *What worked worse than expected?*
- *What would you try differently?*

The psychological safety rules:

- *The team leader shares first, including failures*
- *No output from the experiment session is evaluated for quality*
- *All use cases generated in the experiments are recorded, credited to the person who found them, and fed back into the adoption plan*

Why it works:

- *Autonomy removes the mandate dynamic that creates resistance*
- *Protected time removes the prompt burnout trap of extra work.*
- *Visible leader experimentation normalizes the incompetence stage.*
- *Credit for use case ideas creates investment in the program's success.*

Figure 9.6 - AI Experiment Framework

Celebrating Human Judgment, Not Just AI Efficiency

The narrative of AI adoption in most organizations is relentlessly efficiency-focused. Time saved. Processes automated. Administrative burden reduced. These are real and valuable outcomes. They are also, for many HR professionals, a constant reminder of what is being optimized away.

The change management narrative that sustains adoption over time is not efficiency. It is capability. It is the stories of what HR professionals did with the time that AI gave back to them. The complex employee relations case that was handled better because there was time to think. The manager conversation that went deeper because the administrative preparation was done. The workforce planning insight that emerged because someone had time to analyze rather than just report.

Every AI adoption communication plan should include these stories. Not as an afterthought: as the centerpiece. And alongside the stories of time saved, include the stories of catches: the moments when a human reviewer caught something the AI got wrong, and the outcome was better because a person was paying attention. Celebrating the catch is as important as celebrating the efficiency gain. It reinforces that the human is the accountable party, and that accountability is a feature, not a limitation.

AI adoption succeeds when people can see that the technology is making them more human, not less. That is the narrative that sustains engagement. And it is the narrative that is most consistently absent from HR technology change programs.

Cross-Industry Case Study: Housing Association

CROSS-INDUSTRY CASE STUDY

Organization: A housing association, HR team of 6,900 employees

Function: HR operations and employee relations

The problem:

Following the introduction of an AI document drafting tool, designed to accelerate the production of letters, policies, and employee communications, 70% of the HR team had effectively stopped using it within three months of launch. The team's feedback was consistent and specific: the tool produced outputs that felt generic, that required significant editing to reflect the organization's culture and tone, and that, when distributed in their early form, had attracted negative comments from managers about the quality of HR communications.

The HR Director had mandated the use of the tool as part of a cost reduction initiative. The mandate had created compliance without adoption: team members opened the tool, generated outputs they did not use, and continued writing their documents manually. The tool's usage statistics reported high engagement. The team's actual experience was frustration.

What they did:

The HR Director acknowledged publicly to the team that the mandatory approach had not worked, and that the feedback about output quality was legitimate. She suspended the mandate.

In its place, she introduced a voluntary AI experiment program:

- Each team member was given one hour per week of protected time to experiment with AI on a task of their own choosing. And, not the document drafting tool specifically, but any AI tool for any real work task.

Results were shared in a weekly 10-minute team standup. No evaluation of output quality. The only questions were: what did you try, what worked, what didn't, what would you do differently?

The HR Director shared her own experiments first each week, including the ones that produced worse outputs than she could have achieved manually.

Outcome:

83% voluntary adoption within 4 months, significantly higher than the 30% actual usage achieved under the mandate

The team generated 14 new AI use-case ideas that the original implementation team had not identified, including an ER case summary generator that became the team's highest-value AI application

The document drafting tool was reintroduced six weeks into the experiment program, this time with team members adapting it for their specific contexts rather than using it as prescribed

The key insight:

Autonomy is the antidote to AI anxiety. When people feel they are using the tool rather than being used by it, the resistance evaporates. The same tool that 70% of the team had rejected under mandate was adopted by 83% when the conditions for experimentation were psychological safety rather than compliance. The decisive variable was not the tool. It was never the tool.

Figure 9.7 - Cross-Industry Case Study

GREY AREA DECISION FRAMEWORK

Grey Area: Mandated AI Use

When an organization mandates AI tool usage, it creates a legal and ethical tension that HR leaders are uniquely positioned to navigate, and uniquely exposed to if they do not.

The core question:

Can an employee reasonably refuse to use an AI tool on grounds of conscience, disability, or data privacy concern? The honest answer: it depends on the jurisdiction, the nature of the tool, the specific concern, and whether a reasonable alternative exists. This is not a situation that calls for a blanket policy. It calls for individual assessment.

The framework for HR leaders:

Before mandating the use of any AI tool, ask: Is there a reasonable adjustment that would allow this employee to achieve the same outcome without this specific tool? Does the tool process employee data in a way that requires explicit consent under applicable data protection law? Does using this tool require the employee to do something that conflicts with a sincerely held ethical position?

The governance rule:

Mandating outcomes is defensible. Mandating tools is more complex. Requiring a structured interview question set for every role is a legitimate performance expectation. Requiring a specific AI tool to produce it is more legally and ethically complex in many jurisdictions.

When you receive a refusal:

Step 1: Listen. Document the specific concern.

Step 2: Assess whether a reasonable alternative exists that achieves the same outcome.

Step 3: Consult your legal or data protection advisor before responding.

Step 4: Document every accommodation decision and every outcome.

The documentation rule:

Every refusal, every conversation, every accommodation, and every outcome must be documented. This is your evidence base if challenged, either by the employee who refused, or by a regulator examining your AI governance.

CHAPTER CHECKLIST

☐ I have mapped my key stakeholders on the adoption curve and identified the archetype of each . . . honestly, not optimistically

☐ I have a communication plan for AI adoption that addresses the job security concern directly, before employees have to ask

☐ I have at least one psychological safety mechanism in place for AI experimentation, and I have modeled it myself

☐ I have manager briefing materials ready for the four questions employees will ask . . with specific, honest answers for each

☐ I have a process for addressing AI tool refusals that starts with listening and assessing reasonable alternatives before escalating

PROMPT LIBRARY & WORKBOOK EXERCISES

STAKEHOLDER MAPPING PROMPT

[Help me map my key stakeholders for an AI adoption initiative in an HR context. For each of the following archetypes; Enthusiast, Skeptic, Fearful, Passive, Saboteur, and suggest three specific engagement strategies that are appropriate for an HR professional audience. For each strategy, describe the specific conversation or action, not just the principle.]

COMMUNICATION PROMPT

[Draft an all-staff communication about the introduction of [AI tool] in our HR processes. The communication must: (1) acknowledge the concerns people will have about job security, directly and specifically; (2) explain what the tool does and does not do in plain English; (3) describe the human oversight process for every output that affects an employee; (4) invite questions through a specific channel with a specific response timeline. Tone:

honest and warm, not corporate or enthusiastic. Maximum 300 words.]

WORKBOOK EXERCISE

Think about the last time a change initiative failed in your organization. Write down three specific reasons it failed, and not the official debrief version, the honest one. Now look at your current AI adoption plan and ask: are any of these same factors present? What would you do differently? The patterns of change failure in your organization are usually consistent. This exercise surfaces them before they repeat.

TEAM EXERCISE: THE CONCERNS HARVEST

Give each team member 5 minutes to write down their honest concerns about AI in their specific role, anonymously. Collect the responses. Collate and share the themes in your next team meeting (not the individual responses). Address each concern explicitly in your adoption plan. Then do something specific about at least one concern in the following week, and report back on what you did. This exercise builds trust faster than any communication, because it demonstrates that the concerns were heard and acted on, not just acknowledged.

SKEPTIC'S CORNER FAQ

Q: "AI is going to eliminate jobs. Why would I champion a technology that puts my own people at risk?"

A: This is the right question, and the data is clear. Josh Bersin’s research shows the same outcome across fields like software engineering and healthcare. Jobs did not disappear. They evolved. AI removed the drudgery, not the role. HR’s role is not protection. It is transition. With transparency, reskilling, and dignity. That is the work of this chapter.

Q: “My team is resistant to AI. How do I get them to just use it?”

A: Just" is the problem. Resistance is not obstruction. It is information. It points to trust gaps, anxiety, and unresolved concerns. Treat it as a barrier and it will grow. Diagnose it first. Section 9.2 maps the archetypes and shows you how to meet each one where they are. The strategy follows the diagnosis, not the push to "just use it."

Q: "Our CEO is pushing for faster AI adoption than I think is safe."

A: This is the manage-upward moment, and Chapter 1's Humanly Intelligence Zone Sorter gives you the language to have it. Frame the pace question as a legal and reputational risk management conversation, not a people-are-resistant conversation. "Moving this workflow into Zone 1 without the governance in place exposes us to [specific risk]. The governance takes [specific time]. Here is the sequence that delivers speed without creating exposure." Boards respond to risk language. What they do not respond to is 'our team needs more time to adjust.'

Q: "One team member is actively undermining the AI rollout, what do I do?"

A: Before treating it as a performance issue, treat it as a signal. What specifically are they doing? Is it vocal opposition, passive non-use, or active disruption of others' adoption? Each is a different situation requiring a different response. What is the specific concern underneath the behavior? Is it about the tool, the process, how the rollout was communicated, or something else entirely? Answer those questions before reaching for the performance framework. Most apparent saboteurs are failed communicators, people whose legitimate concerns were not heard early enough and who found other ways to make themselves heard.

What Keisha Does Differently

Keisha does not abandon the tool. She does not pretend the rollout went well when it did not. She does not double down on the mandate that created the resistance.

She holds a team meeting. She acknowledges that the implementation was not working as she had hoped. She asks, specifically and without defensiveness, what would need to be different for this tool to be genuinely useful rather than genuinely frustrating. She listens without interrupting. She does not defend the vendor's promise against her team's lived experience.

Then she suspends the mandate. She introduces the experiment framework. She shares her own clumsy, unsuccessful prompt attempts in the first standup. She credits the team member who identifies the first genuinely useful use case. She reports to the executive team on what she learned, what she is doing differently, and what the timeline for sustainable adoption now looks like.

The room she was losing comes back, and not because the tool improved, but because the leadership did.

Change/Transformation – *42% of companies abandoned AI initiatives in 2025, up from 17% the prior year. Meanwhile, 31% of workers admit to actively undermining company AI efforts – refusing tools, inputting poor data, or slow-rolling projects.* '

~S&P Global Market Intelligence / Writer/Workplace Intelligence, 2025

Chapter 10: Proving the Value: ROI, Measurement, and the C-Suite Conversation

CASE IN POINT

David is the CHRO at a global retail group. Last year, he navigated a complex procurement process to implement an AI talent platform. The price tag: $340,000. At the time, the investment was framed as a "*necessary leap into the future of work.*" The board approved it, albeit with a raised eyebrow from the CFO, Sarah.

Figure 10.1 - David's Story

Now, twelve months later, David is preparing for the annual budget review. He knows instinctively that the platform is working. His recruiters seem less stressed. Time-to-fill for store manager positions feels faster. He has heard anecdotal praise about the caliber of new hires.

But Sarah does not want to hear that recruiters "seem" less stressed. She wants the dollar value of the hours reclaimed. She does not want to hear that hiring "feels" faster. She wants to know how many days of vacancy-related revenue loss were prevented. David spent $340,000 of the company's capital, and he cannot prove any of it with a hard number.

If David walks out of that budget review without a credible ROI story, his next request . . . AI expansion into Learning and Development, will be dead on arrival.

***

There is a particular kind of professional silence that descends on a room when a CFO asks a simple question and no one has a simple answer. HR leaders know that silence well. They have sat in it more than once. They have felt the moment where belief becomes insufficient, where intuition runs out and the only currency that matters is data.

AI investment has raised the stakes of that moment considerably. When your organization commits six or seven figures to an AI platform, in recruiting, in learning and development, in employee relations, in workforce planning, the C-suite is no longer asking whether HR has a seat at the table. They are asking whether HR can justify keeping it.

This chapter is your answer to that question. Not a defensive one. A compelling one. It gives you the measurement framework, the financial translation tools, the ROI calculation methodology, and the boardroom language to make the case for AI investment with confidence, precision, and intellectual honesty. It also gives you the Grey HR Zone, because measurement is never purely technical. It requires judgment, ethics, and the courage to say what you cannot yet prove.

HR leaders who cannot measure the impact of AI investment will lose the budget conversation. HR leaders who can measure it and communicate it in the language of business will define it.

The HR AI Measurement Framework

Measuring AI return on investment in HR is not the same as measuring technology ROI elsewhere in the organization. In a manufacturing setting, AI ROI is calculated by units produced per hour. In HR, the unit is a human life, a career, or a cultural shift. The lag time between an AI intervention, a better matching algorithm, a smarter onboarding sequence, and a downstream business outcome can be months or even years. That complexity does not excuse measurement avoidance. It demands measurement sophistication.

The outcomes HR cares about, retention, culture, engagement, quality of hire, leadership effectiveness, are shaped by dozens of variables simultaneously. AI is one of them. Isolating its specific contribution is legitimately hard. The frameworks in this section are designed to help you do it honestly.

The Three Measurement Categories

Every HR AI initiative should be evaluated across three categories. Think of them as the measurement triangle: efficiency, effectiveness, and ethics.

Efficiency asks: Did AI save us time and money? These are the easiest metrics to quantify and the ones your CFO will ask about first. Time-to-hire. Cost-per-hire. HR team hours reallocated from administrative work to strategic work. They are the foundation of your ROI case.

Effectiveness asks: Did AI improve the quality of our outcomes? This is harder to measure and more important. Quality-of-hire. Skills gap closure rate. Reduction in repeat employee relations cases. Performance improvement in AI-supported roles. These metrics tell you whether AI made HR better, not just faster.

Ethics asks: Did AI reduce our risk exposure? This category is chronically undervalued in ROI conversations because the value lives in things that did not happen. Compliance incidents avoided. Discrimination claims prevented. Escalations that resolved before they became legal matters. The cost of what did not happen is real money. It belongs in your business case.

Leading Indicators vs. Lagging Indicators

One of the most common measurement mistakes HR leaders make is measuring only lagging indicators: metrics that reflect outcomes already achieved. Turnover reduction is a lagging indicator. So is improved quality-of-hire. They are important, but they arrive late. By the time you can prove them, another budget cycle has already started.

Leading indicators tell you whether your AI initiative is on track before the lagging outcomes materialize. Candidate pipeline quality scores. Employee engagement pulse trends. Skills assessment completion rates. Manager confidence scores in AI-supported decisions. These are the metrics that let you walk into a quarterly review and say: the evidence suggests we are heading in the right direction, and here is why.

A robust measurement plan includes both. Lead with leading indicators for ongoing performance reviews. Anchor your annual case to lagging outcomes. Together, they tell a complete story.

The Retrospective Measurement Problem and How to Solve It

Here is the most uncomfortable truth in this chapter: the majority of organizations that have already implemented AI tools did not establish baseline metrics before launch. If that includes you, you are not alone. You are also not without options.

Industry benchmarks serve as credible proxy baselines. SHRM, Gartner, and Deloitte publish annual benchmarks for time-to-hire, cost-per-hire, turnover rates, and HR team productivity by sector and organization size. If you can show that your post-implementation metrics compare favorably to industry benchmarks for organizations at your pre-AI stage, you have a defensible retrospective baseline.

Your HRIS almost certainly holds historical data you have not yet interrogated. Pre-implementation hiring cycle times, case volume trends, training completion rates: the data exists. It requires a deliberate effort to surface and organize it. You can also run a control group comparison: measure a business unit or department that has not yet adopted the AI tool against one that has. The differential tells its own story. A retrospective measurement built on these sources is less precise than a prospective one. It is far more credible than no measurement at all.

Measuring AI Impact Across HR Functions

Different HR functions require different measurement approaches. The following framework organizes key metrics by function, connects them to financial proxies, and translates each into the business line impact language your operational leaders and CFO will recognize.

HR Function	Key Metrics	Financial Proxy	Business Line Impact
Recruitment	Time-to-hire, cost-per-hire, quality-of-hire index, offer acceptance rate, candidate NPS	Recruitment cost avoidance; revenue impact of faster role fills; bad hire cost reduction	Vacancy days reduced; revenue protected per unfilled role; manager time saved on repeat interviews
Employee Relations	Case resolution time, repeat case rate, policy query volume, escalation rate	Legal cost avoidance; HR capacity reallocation; compliance incident reduction	Manager time reclaimed from ER escalations; reduction in productivity loss from unresolved team conflict
L&D	Completion rates, skills gap closure, time-to-productivity for new hires, time-to-full-performance, 30/60/90-day manager readiness ratings, performance correlation post-learning	Training cost per head reduction; productivity gains; retention improvement	Days to quota attainment for sales roles; output per new hire at 30/60/90 days; error rate reduction in technical roles; manager-rated team readiness scores; reduction in repeat performance conversations
OD / Workforce Planning	Engagement delta, retention improvement, change adoption rate, succession fill rate, internal mobility rate	Turnover cost avoided; bench strength improvement; change initiative ROI	Business unit performance correlation to engagement scores; critical role vacancy reduction; leadership pipeline depth
Agentic AI	Error rate, escalation rate, time saved per workflow, compliance incident rate	Staff cost avoidance; risk reduction value; process capacity gain	HR transactions processed without human intervention; SLA compliance rate; administrative error reduction

Figure 10.2 - The HR AI Measurement Framework - From Function Metrics to Business Line Impact

Recruitment: The Most Visible ROI

Recruitment is where most organizations see their first measurable AI returns, and for good reason. The inputs and outputs are relatively discrete. You post a role. You receive applications. You screen, interview, assess, extend an offer, and onboard. AI touches multiple stages of that process, and each stage has a measurable cost and time value.

Time-to-hire improvement is the entry point. If AI screening reduces your average time-to-hire from 42 days to 28 days, and your average unfilled role costs the business $1,200 per day in lost productivity, a figure you can calculate from total role compensation divided by working days, the math is straightforward. Fourteen days recovered per hire. Multiplied by hire volume. That is a real number your CFO can verify.

Quality-of-hire is the harder and more important metric. Define it before you measure it. A quality hire, for your organization, might be: still employed at 12 months, performing at or above target in their first performance review, and rated positively by their hiring manager at the 90-day mark. Build a composite quality-of-hire index from those three data points. Track it before and after AI screening integration. The delta is your quality argument.

Employee Relations: Measuring What Did Not Escalate

Employee relations is the function where AI's risk reduction value is most significant and least often measured. An AI-assisted case management system that helps ER professionals identify policy-compliant responses, flag escalation risk early, and document consistently reduces both the frequency and severity of cases that become formal complaints or legal actions.

The financial value of that reduction is calculable. The average cost of defending an employment discrimination claim in the United States, before settlement or judgment, runs from $75,000 to $125,000 in legal fees alone. A single prevented escalation to litigation pays for significant AI investment. Your ER leaders know which cases were close calls. That institutional knowledge, systematically captured, is your ethics ROI data.

L&D: Connecting Learning to Performance

The chronic measurement challenge in learning and development is the attribution gap. Learning happens. Behavior may change. Performance may improve. Proving that the learning caused the improvement is where most L&D measurement frameworks break down.

AI-powered learning platforms offer a structural advantage here: granular data on skill development over time. If your platform tracks individual learner progress against a skills taxonomy, and you can correlate skills gap closure with 90-day performance scores, you have a defensible L&D ROI argument that most organizations cannot make. Use it. The time-to-productivity metric for new hires is particularly powerful: reduce the ramp period by two weeks, multiply by hire volume and daily productivity value, and you have a number that resonates in any budget conversation.

The CFO Conversation: Translating HR Metrics into Business Language

Here is the hard truth about most HR AI business cases: they are written in HR language and read by people who speak finance. That translation gap is where good initiatives go to die.

A CFO does not wake up thinking about time-to-hire. They wake up thinking about three things: cost reduction, risk reduction, and revenue impact. Every HR metric you present needs to connect, explicitly and credibly, to one of those three categories. If it does not, it is a data point. It is not an argument.

The Three-Number Framework

Train yourself to present every AI metric through a three-number lens. Abstract metrics do not move budgets. Translated metrics do.

- **Cost reduction:** What did AI allow us to spend less on? The translation looks like this: "By using AI to automate interview scheduling, we reclaimed 4,000 hours of recruiter time, which allowed us to absorb growth without adding headcount, saving $280,000 in annual salary costs." That is a sentence a CFO can work with. "We improved recruiter efficiency" is not.

- **Risk reduction:** What bad outcomes did AI help us avoid? Translate it: "Our AI compliance tool identifies high-risk payroll errors before each run. This reduces our exposure to a Department of Labor audit, where the average fine for an organization of our size is $150,000 per incident." CFOs understand risk-adjusted value. Use that language.
- **Revenue impact:** How did AI help the business perform better? The most powerful version of this argument ties directly to operations: "AI-driven talent matching reduced store manager vacancy days by an average of 12 days per opening. Based on average daily store revenue, this equates to $1.2M in protected sales that would otherwise have been lost." Revenue impact is the hardest HR metric to credibly claim. Build it carefully. Claim it conservatively. A modest, defensible number is worth more than an ambitious one that gets challenged.

The Risk Reduction Argument: Putting a Price on the What If

CFOs often view HR as an insurance policy. Your job is to make the premium visible. Use published cost data to build the case.

A bad hire costs between 1.5x and 2x the role's annual salary. If AI improves quality-of-hire and prevents just five bad hires in a year at an average salary of $80,000, you have saved the company between $600,000 and $800,000. The average cost to defend an employment discrimination claim, before settlement or reputational damage, is roughly $160,000 in legal fees alone. If AI auditing makes fair hiring more systematic, it functions as a significant financial shield. These are not hypotheticals. They are published figures your legal and finance teams will recognize.

Building the One-Page Board Summary

Boards make better decisions with less information, not more. The HR AI board summary fits on one page. Give them four quadrants, nothing more.

- **Financial Impact**: Total ROI, current and projected. Three-scenario model if the data warrants it.
- **Operational Efficiency**: Reclaimed hours, process acceleration, headcount capacity gained without headcount added.
- **Strategic Value**: How AI is advancing the organization's primary strategic goal. Connect it explicitly: "This initiative is enabling rapid scaling into two new markets without proportional HR staff growth."
- **Governance and Risk**: Ethics audit results. Bias review status. Data security posture. Boards increasingly expect this quadrant. HR leaders who lead with it earn trust they would not otherwise have.

What the summary should not contain: methodology footnotes, technical platform descriptions, percentage improvements without financial translation, or hedged language that signals uncertainty about your own numbers. If you are uncertain about a number, use a range. A range is a position. Vagueness is the absence of one.

The ROI Scenario Model

Present three scenarios: conservative, moderate, and optimistic. Each should clearly state its assumptions.
A three-scenario model does something important: it demonstrates analytical rigor and simultaneously manages expectation in both directions. It signals that you are not overselling the platform and that you have thought seriously about the variables that affect outcomes.

The following illustrative model shows how a three-scenario framework might look for a mid-market organization for twelve months into an AI talent platform implementation:

Assumption	Conservative	Moderate	Optimistic
Time-to-hire improvement	10%	20%	35%
Cost-per-hire reduction	$800	$1,400	$2,100
Quality-of-hire improvement	5%	12%	20%
Turnover reduction (attributable)	3%	7%	12%
HR team hours reallocated/yr	200 hours	480 hours	800 hours
Estimated annual value	$180K	$420K	$780K
Estimated 3-year ROI	1.6x	3.7x	6.9x

Figure 10.3 - ROI Scenario Model

Note: Values above are illustrative. Your actual model should use your organization's cost-per-hire baseline, average daily productivity value per role, and documented pre-implementation benchmarks.

GREY AREA DECISION FRAMEWORK

Grey Area: Attributing Business Outcomes to AI

The hardest measurement problem in this chapter is not calculating the numbers. It is knowing how much of the credit AI deserves. When multiple variables change at the same time, market conditions shift, your recruitment team reorganizes, your employer brand investment pays off, proving that AI caused a specific improvement is genuinely difficult. This is not a flaw in your measurement approach. It is an accurate reflection of how organizations actually work.

The Grey Zone principle here is attribution honesty. Three rules govern it:

- **Rule 1:** Never overstate AI attribution. A single challenged claim destroys the credibility of your entire measurement case. One CFO who catches an inflated number will discount everything else you present.

- **Rule 2:** Document confounding variables as rigorously as you document outcomes. This is your intellectual honesty defense. It demonstrates analytical sophistication and signals that you are not hiding inconvenient context.

- **Rule 3:** Use contributing factor language. AI was a contributing factor. Here is the correlation data. Here is what else changed. Here is our best estimate of AI's specific contribution. That is more defensible than AI produced this outcome, and boards find it more credible.

The safer framing in every board conversation: AI enabled our team to do X, which contributed to outcome Y. That framing is both honest and compelling. It does not concede AI's value. It contextualizes it.

Building a Continuous Measurement Culture

The single biggest measurement mistake organizations make is treating AI ROI as a one-time calculation. They implement a platform, measure impact at the twelve-month mark, present to the board, and move on. This approach systematically understates AI value, and it leaves organizations poorly positioned for the investment decisions that follow.

Unlike a piece of machinery that depreciates over time, AI software often appreciates in value as it learns your organization's specific patterns and nuances. A screening algorithm that improves marginally in Year 1 improves substantially in Year 2. A skills taxonomy that is incomplete at launch becomes a strategic asset eighteen months later. A platform might show minimal measurable ROI in the first six months while the team is still adopting it, 20% return in the following six, and 150% or more by Year 2 as the model matures. Point-in-time measurement captures only the early portion of that curve. Continuous measurement captures the full arc.

The Quarterly AI Performance Review

Build a simple operating rhythm: a 60-minute quarterly AI Performance Review with the HR leadership team. The agenda is consistent every quarter, which means it is low-friction to prepare and high-value to run. Three questions anchor it: What has changed since last quarter? Are the tools providing actual insight, or just more data for the team to wade through, the signal-to-noise question every AI investment eventually has to answer? And what is the user adoption rate? If the team is not using the tool, the ROI is zero, regardless of what the vendor dashboard says.

This operating rhythm does three things simultaneously. It normalizes measurement as a leadership practice, not an annual event. It creates a structured data trail that makes board presentations dramatically easier to prepare. And it positions HR leadership as the kind of function that proactively manages its investments rather than defending them reactively when questioned.

When to Stop, Pivot, or Double Down

Not every AI initiative will deliver. Some will underperform. Some will fail. The ability to recognize underperformance early and respond with clarity rather than defensiveness is one of the hallmarks of sophisticated AI governance. The following decision framework structures that judgment:

Signal	What It Means	Recommended Action
Metrics trending up, team aligned	AI is working as intended	Double down. Scale the initiative. Document and share the playbook.
Mixed results, unclear attribution	Normal. AI rarely works perfectly in Year 1.	Pivot. Refine the use case, improve data quality, and adjust the workflow.
Metrics flat or declining, team disengaged	The tool is not solving the right problem, or adoption has failed.	Stop or pause. Conduct a root cause review before reinvesting.
Results strong, but ethics concerns raised	AI may be introducing bias or equity risk.	Pause. Audit the model. Do not optimize for efficiency at the cost of fairness.

Figure 10.4 - Decision Framework

The ethics row in that framework deserves particular emphasis. An AI initiative that is delivering efficiency gains but raising bias concerns is not a success story. It is a Grey HR Zone that requires immediate attention. The operational case for the tool does not outweigh the ethical obligation to address systemic fairness risk. HR's credibility as the guardian of people-centered practice depends on holding that line, even when it is financially inconvenient.

Cross-Industry Case Study: US Healthcare Network

Cross-Industry Case Study: US Healthcare Network

Situation

A large US healthcare network's Chief People Officer was facing board scrutiny after 18 months of a $2.1M AI HR platform investment — with no formal ROI measurement in place. The platform had been implemented during a period of significant organizational change, making clean attribution inherently complex.

What They Did

Rather than waiting for the board presentation, the CPO engaged an internal analytics team to retrospectively build a measurement framework using pre-implementation baseline data from the HRIS. Industry benchmarks were used to supplement gaps in the historical data. The measurement team explicitly documented confounding variables, a system-wide merger, and a leadership restructuring, alongside the outcome data.

Measured Outcomes

Time-to-hire reduction: 31% improvement, valued at approximately $1.1M in unfilled role cost avoidance across a high-volume clinical hiring function.

Turnover cost avoided: A 4.2% improvement in 12-month retention among AI-screened hires, valued at $1.8M, using the organization's standard turnover cost model.

Compliance incident reduction: A 28% reduction in hiring-related compliance escalations, valued at $900K using average external legal cost per incident.

HR team time reallocation: Approximately 800 hours per year redirected from administrative screening tasks to strategic talent advisory work, valued at $700K in staff capacity.

Total Calculated Value: $4.5M against a $2.1M investment over 18 months.

Outcome

Full board approval for platform renewal and expansion into two additional business units. The CPO was promoted to the Executive Committee within six months of the presentation. The board noted that the quality of the measurement case, including its explicit documentation of confounding variables and conservative scenario assumptions, was a significant factor in the decision.

Figure 10.5 - Cross-Industry Case Study

Chapter Checklist

☐ I have **established baseline metrics**, internal or benchmark, for every AI initiative currently running in my organization.

☐ I have a **measurement plan** with at least two leading indicators and two lagging indicators for each initiative.

☐ I have translated at least **one HR AI metric into a financial number** my CFO would recognize and find credible.

☐ I have identified the **risk reduction value** of at least **one AI initiative**, using published cost benchmarks where needed.

☐ I have a **quarterly AI performance review** scheduled as a recurring leadership team meeting.

☐ I have completed or started a **one-page board summary draft** for my organization's largest AI investment.

☐ I have reviewed my **measurement case for attribution** honesty and can articulate what AI contributed versus what other factors contributed.

Prompt Library and Workbook Exercises

ROI Calculation Prompt

[Help me calculate the ROI of our AI [recruitment / L&D / ER] tool. Pre-implementation baseline: [describe what you were measuring before]. Post-implementation data: [describe current metrics]. Additional variables that changed during this period: [describe]. Build me a conservative, moderate, and optimistic ROI scenario with the assumptions clearly stated for each, and flag which assumptions are most sensitive to challenge.]

CFO Language Translation Prompt

[Translate these HR AI outcomes into language a CFO would find compelling: [list your outcomes]. For each outcome, suggest the financial proxy I should use, the data I need to collect, the calculation methodology, and the most defensible way to present it to a board audience that will push back on attribution.]

Board Summary Prompt

Draft a one-page board summary for the following AI HR investment: [describe platform, investment level, timeframe]. Include: investment summary, measured outcomes in financial terms, key assumptions, confounding variables we have documented, and a recommendation for next steps. Tone should be confident, precise, and appropriately hedged where evidence is correlation rather than causation.

Workbook Exercise 1: The Measurement Deficit

Select your single largest AI investment. Write down three things: (1) what the vendor promised it would deliver, (2) what you can actually prove it has delivered today, and (3) the gap between the two. That gap is your measurement deficit, and your priority for the next quarter. If you cannot measure the promise, you must build the sensor.

Workbook Exercise 2: The Red/Amber/Green Audit

Run a measurement readiness audit with your HR leadership team. For each active AI initiative, ask: "If our CFO asked for proof of ROI tomorrow, what could we show them?" Score each initiative red (no measurement case), amber (partial), or green (ready to present). Every red initiative needs a measurement plan before the next budget cycle. This is not optional. It is risk management.

Skeptic's Corner FAQ

Q: "We did not collect baseline data before implementing. Can we still measure ROI?"

It is never too late, but you will have to work harder. Use industry benchmarks from SHRM or Gartner as a proxy baseline. You can also use a control group approach: measure a department that has not yet adopted the tool against one that has. The differential tells a credible story without requiring historical data you did not collect. Retrospective measurement is less robust than prospective measurement. It is infinitely better than silence.

Q: "Our AI investment has not delivered the promised returns. How do I tell the board?"

Be the first one to say it. Boards respect data-driven honesty far more than optimistic spin. Use the Section 10.4 decision framework to explain why performance has fallen short, poor adoption, weak integration, misaligned use case, and present a clear plan to pivot or wind down. This approach builds the trust that gets your next proposal approved.

Q: "How do I value risk reduction when nothing bad has happened yet?"

You do not wait for a fire to value a fire extinguisher. Use the actuarial approach: research the average settlement for an Equal Employment Opportunity Commission claim, or the average cost of a GDPR breach for an organization of your size. Present these as financial exposure that AI is actively mitigating. CFOs understand risk-adjusted value perfectly. It is the same methodology used by every insurance underwriter and enterprise risk function in business. HR is entitled to the same framework.

Q: "What if our CFO simply does not believe the numbers?"

A CFO who challenges your methodology is doing their job. Prepare for it. Know which assumptions are most sensitive. Know which data sources are most credible. Know which numbers are conservative estimates and which are actuals. The challenge is not a rejection of your case. It is an invitation to demonstrate the rigor behind it. The HR leaders who have done that rigor welcome the challenge. The ones who have not dread it.

The Currency of Credibility

David has three weeks. He has $340,000 invested, a CFO named Sarah who asks the one question nobody can answer, and a budget review that will determine what comes next, including whether his L&D AI proposal ever sees daylight.

With the framework in this chapter, David can **build a retrospective measurement case**, translate his **outcomes into financial language**, construct a **three-scenario ROI model**, and walk into that room with a **board summary** that is precise, honest, and defensible. He can show what AI delivered. He can show what else changed. He can show what he recommends next and why.

That is not spin. That is leadership. The measurement capability this chapter describes is not a compliance exercise. It is the difference between an HR leader who asks for budget and one who argues for it. Between a CHRO who is viewed as an HR administrator and one who is viewed as a business leader who happens to run People.

The organizations that build this capability now will be the ones whose AI investments compound. The ones that do not will keep having the wrong version of the same conversation, intuitive, unquantified, and ultimately unconvincing.

You now have what you need to make it a different one.

The **Grey HR Zone** in measurement is not about the numbers. It is about the honesty and judgment you bring to them. Lead with both.

Worth Remembering

The C-suite is not asking whether AI is interesting. They are asking whether it is working.

> **AI's ROI –** *88% of HR leaders say their organizations have not realized significant business value from AI tools. Only 1 in 5 AI investments delivers any measurable ROI at all. The conversation HR needs to be ready for is not about adoption. It is about proof.*
>
> ~ Gartner, October 2025

Chapter 11: Data Privacy, Compliance, and the Legal andscape of AI in HR

CASE IN POINT

The Moment It All Unravels

Elena is a Global HR Director overseeing teams in 14 countries. Three months ago, her analytics team rolled out an AI-powered sentiment tool. It analyzed Slack messages, emails, and meeting transcripts to predict burnout and flight risk.

Figure 11.1 - Elena's Story

The pilot? A huge success. The insights? Impressive. Leadership? Thrilled.

Then came the meeting.

Her Data Protection Officer didn't waste time:

- In Germany, the Works Council was threatening legal action

- In France, the tool violated employee privacy protections
- In the EU, it triggered GDPR Article 22 concerns
- Under the EU AI Act, it likely qualified as high-risk AI

And here was the real issue:
No DPIA. No legal review. No employee transparency.

Elena wasn't asked before it launched. But she's the one responsible for fixing it.

This isn't a rare situation. It's one of the most common failure patterns in AI adoption:

- Technology moves fast.
- Governance shows up late.
- HR gets handed the cleanup.

This chapter exists so you don't become Elena.

***

The legal and regulatory environment around AI in HR is evolving faster than most organizations are prepared for, and that gap is exactly where risk lives.

Let's be clear: this is not about turning HR leaders into lawyers. This is about making sure you are **informed enough to ask the right questions**, **stop risky decisions** before they happen, and **build systems** that won't collapse under *regulatory pressure* later.

Because here's the truth most organizations learn too late: AI doesn't create compliance risk. Uninformed adoption does.

Most of this chapter is about what the law requires. But one compliance question sits underneath all of it, and most HR leaders never think to ask it. Who controls the environment your AI runs in? Not who sold it to you. Not who supports it. Who owns the data, the access, and the audit trail. The regulatory frameworks ahead tell you what the rules are. The section titled ***The Architecture Question Nobody Is Asking*** tells you why the environment your AI lives in may matter more than the contract that governs it.

Alan Landever, this book's technical contributor, addresses that architecture question directly. It is worth keeping in mind as you move through everything that comes before it.

The Regulatory Landscape – What HR Leaders Actually Need to Know

The waiting and watching era is over.

Regulators are no longer observing AI in HR. They are actively governing it. And HR sits directly in the blast radius, because AI decisions affect hiring, performance, pay, termination, and employee monitoring. In other words: people's livelihoods.

The EU AI Act: The Global Benchmark

The EU AI Act is the most important regulation to understand, even if you are not based in Europe. Why? Because it applies based on where your employees are, not where your company is headquartered.

Most HR AI tools fall into high-risk categories, including:

- Recruitment and candidate screening
- Performance evaluation and rating
- Promotion and pay decisions
- Employee monitoring and behavior analysis

If your tool touches any of those areas, you are in scope.

What that means in practice: you are now responsible for human oversight, meaning AI cannot make final decisions alone; transparency, meaning employees must know AI is being used and how; documentation, meaning you must be able to demonstrate how the system works; and risk management including bias monitoring.

This is enforceable. The time to build compliance is now, not when a regulator calls.

GDPR: The Rule Everyone Thinks They Understand But Usually Does Not

GDPR remains the backbone of data protection in Europe, and it matters more in the age of AI, not less.

Article 22, Automated Decision-Making: employees cannot be subject to decisions made solely by automated processing that significantly affect them. If your AI screens candidates, flags performance risks, or influences promotion decisions, you must have meaningful human involvement. Not a rubber stamp. Actual judgment.

Articles 13 and 14, Transparency: employees must be informed about how their data is being used, including when AI is involved. Privacy notice language buried in an onboarding pack does not satisfy this obligation. Transparency must be meaningful and in plain language.

Article 35, Data Protection Impact Assessment: processing likely to result in high risk to individuals triggers a mandatory DPIA. AI tools that profile employees, monitor behavior, or contribute to significant individual decisions will almost always meet this threshold.

The US: The Patchwork Reality

There is no single AI law in the United States. Instead, HR leaders are navigating an expanding web of state regulations that adds new requirements every legislative cycle.

The takeaway: you are probably already regulated. You just do not realize it yet.

UK and APAC: Diverging, But Catching Up

The UK is moving toward a more flexible, innovation-driven model but remains grounded in GDPR-equivalent principles. Track UK ICO guidance as your primary compliance reference for UK employees, and do not assume EU and UK obligations are identical.

APAC is rapidly formalizing. Singapore's PDPA and Model AI Governance Framework are influential across the region. Australia's Privacy Act reforms are progressing. China's AI regulations apply to companies operating in-market. And multiple APAC jurisdictions now enforce data localization requirements, meaning your AI tool's cloud infrastructure location matters, not just its function.

Your AI tool might be compliant in one region and illegal in another.

The Direction of Travel

Every signal points the same way: mandatory transparency, mandatory bias auditing, mandatory human oversight, and less tolerance for opaque systems.

The organizations that get ahead of this will not be the ones reacting to each new regulation as it lands. They will be the ones that built compliance into their operating model before they needed it.

The HR AI Data Privacy Audit

Start with an honest question: what employee data are our AI tools collecting right now?

Most HR leaders cannot answer it. That is not a failure. It is the current reality. But it is also your biggest compliance exposure. The audit is how you find out.

Step 1: Map Your Data – No Shortcuts

List every AI tool currently in use across your HR function. This includes AI embedded in your HRIS, standalone recruitment or performance platforms, learning systems, productivity tools with AI features, and any pilot or experimental tools not yet formally approved.

For each tool, answer four questions: What employee data does it collect? Where is that data stored? Who can access it? How long is it retained?

Any answer of I do not know is an immediate priority, not next quarter's agenda.

Step 2: The Data Minimization Reality Check

AI vendors love data. Regulations do not. The data minimization principle requires that you collect only what is adequate, relevant, and necessary for the stated purpose.

The vendor problem: AI tools frequently collect more than their primary function requires, including behavioral patterns, sentiment signals, and usage metadata, because that data has commercial value to the vendor. Watch especially for contract provisions allowing vendors to use your employees' data to train or improve their models. If that is happening without explicit written agreement, that is a significant due diligence flag.

Step 3: Consent vs. Transparency – The Uncomfortable Truth

In most employment contexts, employee consent does not legally hold up the way organizations assume. GDPR's requirement for freely given consent is difficult to satisfy in a relationship where the power imbalance is inherent.

Which means your obligation shifts to radical transparency. Employees need to understand what AI is being used, what data it uses, and how it impacts decisions affecting them. Not buried in legal language. Actually understandable.

Step 4: Cross-Border Data – The Hidden Risk

If your AI tool processes EU employee data on infrastructure located outside the EU, you need proper legal mechanisms in place: Standard Contractual Clauses, Binding Corporate Rules, or reliance on an adequacy decision for the destination country. Most organizations assume this is handled by the vendor. It usually is not confirmed.

Step 5: Data Retention – The Silent Liability

AI tools tend to retain data far longer than operational necessity requires, often because that data has value for model training.

GDPR's storage limitation principle requires that data is kept only as long as necessary. Your vendor contracts should specify retention periods, deletion obligations on contract termination, and a verification process. If the answers are vague, fix the contract before the data exposure becomes a regulatory one.

Building Compliance Into AI Procurement

Here is the blunt truth: if you wait until after deployment, you are already behind.

Compliance must be built into procurement, not retrofitted after the contract is signed. Elena's situation was not the result of malice. It was the result of a process that treated compliance as someone else's problem.

The DPIA – Your Best Tool, Not a Bureaucratic Burden

A Data Protection Impact Assessment identifies and mitigates the data protection risks of a new activity before it begins. Under GDPR, it is mandatory when processing is likely to result in high risk, a threshold most HR AI tools will meet.

A DPIA for an HR AI tool should address: the purpose and legal basis for processing; what data is collected and whether that is proportionate; the specific risks to employees including inaccuracy, bias, and misuse of outputs; what human oversight mechanisms are in place; and what rights employees have in relation to automated processing.

You do not need a law degree to run a DPIA. You need a structured process and the right questions. The Prompt Library at the end of this chapter gives you a starting framework. For any tool meeting the six trigger criteria in this section, involve your legal and data protection teams.

The Non-Negotiable Contract Clauses

Every HR AI vendor agreement should contain provisions in these eight areas. If they are absent, add them before signing.

Clause Area	What It Must Cover
Data use rights	Prohibition on using employee data to train or improve vendor models without explicit written consent
Data residency	Where data is stored; restrictions on transfers outside specified jurisdictions
Sub-processors	Full disclosure of any third parties with access to data; notification obligations for changes
Retention and deletion	Specific retention periods; deletion obligations on contract termination; verification process
Security standards	Minimum certifications (ISO 27001 or equivalent); breach notification timelines
Audit rights	Your right to audit vendor compliance, or require third-party audit results
Regulatory change	Process for addressing compliance impact of new legislation affecting the tool
AI Act obligations	For high-risk AI: transparency documentation, human oversight capabilities, bias testing results

Figure 11.2 - Non-Negotiable Contract Clauses

NOTE: If these aren't in the contract – you don't control your risk.

When to Call Legal – The Six Trigger Points

HR leaders should be informed enough to identify compliance risk. They should not be expected to carry it alone. These six situations require legal involvement . . . not a framework, not a checklist:

- The tool is likely to be classified as high-risk AI under the EU AI Act.
- The tool involves automated decision-making with legal or significant effects on employees.
- The tool processes special category data, such as; health, biometric, union membership, or political opinions.
- The tool involves cross-border data transfer to countries without EU adequacy decisions.
- A DPIA has identified high risks that cannot be mitigated through standard measures.
- You have received a data subject access request, a regulator inquiry, or a complaint related to AI use.

Don't wait. Seriously.

The Global Compliance Quick Reference

You don't need to track everything. You need to track what matters.

Region-by-Region: Key Obligations and Red Lines

[illegible]	[illegible]	[illegible]
EU	[illegible]	[illegible]
UK	[illegible]	[illegible]
US	[illegible]	[illegible]
APAC	[illegible]	[illegible]

The Smart Strategy: One Standard, Applied Everywhere

Managing 14 different compliance levels is chaos. The most operationally sustainable approach is to identify your most regulated market and make it your global baseline.

One high standard. Applied consistently. That's clarity and trust.

The Compliance Calendar: Key Triggers for 2025-2027

Trigger / Timing	Action Required
EU AI Act — 2025-2026	Confirm high-risk classification for all HR tools. Implement mandatory human oversight and technical documentation requirements.
Annual — Q1	Review all HR AI vendor contracts for compliance clause currency. Confirm no new sub-processors or data transfer changes have occurred.
Annual — Q2	Run regulatory horizon scan. Assess the impact of new US state legislation on hiring and monitoring tools.
On procurement	Complete the DPIA checklist. Obtain legal review for any tool meeting the six trigger criteria. Confirm employee notification process before launch.
On contract renewal	Renegotiate compliance clauses to the current standard. Obtain updated data processing documentation from the vendor.
On regulatory inquiry	Immediate legal involvement. Preserve all relevant documentation. Do not communicate with a regulator without legal guidance.

Figure 11.5 - Compliance Calendar & Triggers

Building a Monitoring Rhythm Without a Dedicated Compliance Team

This isn't about perfection. It's about not being blindsided. Here's a process that works for teams of any size:

- Identify your three to four highest-exposure jurisdictions. The markets where you have the most employees and the most AI deployment. These are your monitoring priorities.

- Assign ownership by region. Ownership doesn't mean legal expertise. It means tracking two or three reliable sources and sharing updates quarterly.
- Use authoritative sources: the EU's AI Act portal, the UK ICO website, the IAPP for US state law tracking, Singapore PDPC for APAC guidance.
- Meet quarterly to review updates and assess implications for current tools. Not a new meeting, a standing agenda item in your existing operating rhythm.
- Maintain a simple compliance log: a shared document tracking each AI tool, its operating jurisdictions, last review date, and open action items.

The Architecture Question Nobody Is Asking

Most compliance conversations in HR AI stop at the vendor contract. They should not.

The contract governs what the vendor promises. It does not govern what actually happens to your data inside a system you do not control, managed by a company whose business model you may not fully understand.Alan Landever, Founder and CTO of Landever.AI and this book's technical contributor, frames it this way: the single biggest unmanaged compliance risk in HR AI is not the tool itself. It is the environment the tool runs in.

The Tools Your Team Is Already Using

When your HR team uses ChatGPT, Microsoft Copilot, Google Gemini, or AI-embedded features inside platforms like Workday, Microsoft Viva, or Slack, every prompt and every data input is processed by infrastructure you do not own. Every employee name. Every performance note. Every investigation summary. Every compensation detail. In many cases, that data informs the vendor's next model iteration. Your employees' most sensitive information leaves your building and does not come back under your control.

The same applies to any AI tool where you did not negotiate and confirm data handling terms. If you did not ask the question, you do not know the answer.

That is not a technology problem. That is a data sovereignty problem. And data sovereignty is a compliance obligation.

What the Fine Print Actually Says

Microsoft Copilot sits inside the Microsoft 365 ecosystem, which gives many HR leaders a false sense of security. The enterprise licensing agreements do provide stronger data protections than consumer tools, but the data still transits Microsoft's infrastructure, and the governance obligations on your side, knowing what is being processed, where, and by whom, remain yours to manage. Most organizations are not managing them.

ChatGPT, including ChatGPT Enterprise, operates on OpenAI's infrastructure. Even with enterprise agreements that limit training on your data, you are still operating inside an environment you do not control, cannot audit independently, and cannot produce for a regulatory review without going through the vendor.

Google Gemini, integrated across Google Workspace, carries similar considerations. The convenience of having AI embedded in the tools your team already uses is real. So is the compliance exposure when sensitive HR data flows through it without a documented governance framework.

None of this means these tools have no place in HR. They do. The question is which HR tasks they are appropriate for, and which tasks require a controlled environment.

The Governance Alternative

The governance alternative is private AI infrastructure: systems that run on your network, or in a controlled private cloud, where your data is used to generate a response and then stays put. No third-party access. No model training on your employees' records. Full audit visibility. Role-based access controls you define and own.

This matters for HR specifically because the data HR handles sits at the highest sensitivity tier in any organization. Compensation decisions. Investigation records. Medical accommodations. Performance narratives. Termination documentation. None of it should be transiting through an environment you cannot audit, control, or account for in a regulatory review.

The Test Your Current Environment Needs to Pass

The practical question is not whether private infrastructure is better in the abstract. It is whether your current AI environment could survive a data subject access request, a regulatory audit, or a litigation hold. If the answer is uncertain, the environment is the problem, not the policy.

Private AI infrastructure is not a luxury reserved for large enterprises. Accessible, cost-effective private deployment options exist for organizations of most sizes. What it requires is a deliberate governance decision to treat environment control as a compliance requirement, not an IT preference.

Data governance is an enterprise obligation. But HR is the function that understands what is at stake when employee data is mishandled. That makes HR the voice that must drive this conversation, not IT, not procurement. HR, because HR knows what the data means and what it costs when it is compromised.

> **TECHNICAL RESOURCE:** Alan Landever, this book's technical contributor, founded Landever.AI to help organizations implement private, secure AI infrastructure built for real operational use. For HR leaders ready to explore what environment control looks like in practice >>>https://landever.ai/

Cross-Industry Case Study: Mid-Market Professional Services Firm

CROSS-INDUSTRY CASE STUDY

Organization: A mid-market professional services firm, with approximately 4,200 employees across 14 countries, operating across audit, advisory, and tax practices

Function: Global HR Operations

The Problem

The firm's AI-powered performance management tool had been in operation for two years when the EU AI Act's high-risk provisions came into force. An internal compliance review identified the tool as a high-risk AI system. That classification triggered mandatory transparency and human oversight obligations that the firm was not meeting.

What They Did

Rather than treating this as a European problem, the Global HR Operations Director made a deliberate decision to apply the EU AI Act standard globally. The logic was straightforward: maintaining different levels of human oversight, rigorous in Europe and lighter everywhere else, created inconsistency, complexity, and a two-tier standard for how employee performance was being assessed.

The firm conducted a full compliance review. They implemented mandatory human review for all AI-generated performance outputs. No performance rating could move forward without a qualified manager reviewing and signing off on the AI recommendation. They built an employee notification process that explained clearly when AI was contributing to performance assessment, and what that meant for the decisions being made.

The Outcome

Full EU AI Act compliance in 14 weeks. More significantly, the process became the global standard for the organization's HR AI governance, one consistent framework replacing a patchwork of regional variations.

The Lesson

Treating one market's compliance requirement as a global standard simplified governance significantly. The strictest standard became the floor, not the ceiling. And the transparency and human oversight requirements that were legally mandated in Europe turned out to improve the quality and credibility of performance decisions everywhere.

What the case study does not tell you is where that performance tool's data was living during those 14 weeks of remediation. Who had access to it. Whether the vendor could produce an audit trail on demand. Regulatory compliance and environment control are not the same thing. This organization fixed the first problem. Whether they addressed the second is the question every HR leader should be asking before procurement, not after a flag.

THE GREY HR ZONE: Monitoring vs. Surveillance

This is where things get real. Just because something is legal doesn't mean it's right.

AI tools that analyze employee communications, behavior, or sentiment sit at a genuinely contested intersection. On one side: the legitimate value of people analytics, understanding engagement, identifying flight risk, supporting wellbeing. On the other: the line between insight and surveillance, which is both legally variable and ethically significant.

The legal line varies by jurisdiction. The ethical line is more consistent.

Use these **four tests** to navigate the zone:

1. **The Consent Test**: Would your employees be okay with this analysis if they truly understood what was being collected and why? Not the privacy notice version, the plain-English version. If the answer is probably not, that is information.

2. **The Transparency Rule:** If AI analyzes individual employee behavior, disclose it. Even if local law does not yet require it. The absence of a legal requirement is not the same as the absence of an obligation.

3. **The Purpose Limitation Rule:** Data collected for one purpose cannot be repurposed for another without explicit consent. Engagement survey data does not feed performance assessment. Wellbeing tool data does not inform disciplinary decisions. This is both a legal requirement and a trust requirement.

4. **The Global Standard Rule:** Apply the strictest standard in your operating jurisdictions as your global baseline. Two-tier employee standards, that are rigorous in Europe, lighter everywhere else and are both ethically questionable and operationally complex.

Trust is fragile. Once employees feel watched instead of supported, they disengage or they game the system. Either outcome is a people strategy failure , regardless of whether the tool was legally compliant.

The Grey HR Zone position: AI should expand what HR can understand. It must not become a tool for surveillance that employees never agreed to and cannot see.

CHAPTER CHECKLIST

☐ I know which of our AI tools are high-risk under the EU AI Act, and what obligations that classification creates.

☐ I have completed an inventory of every AI tool in our HR function, including what data each collects, where it is stored, and who has access.

☐ I have a DPIA process in place for new AI tool procurement, and I know the six trigger points that require legal involvement.

☐ I have reviewed our AI vendor contracts for the eight compliance clause areas listed in this chapter.

☐ Employees are clearly and plainly informed about when AI is used in decisions that affect them.

☐ Cross-border data transfer risks are assessed and appropriate safeguards are confirmed for each tool.

☐ I have assessed whether the AI environments my team uses could withstand a data subject access request, a regulatory audit, or a litigation hold.

☐ I have a regulatory monitoring process with named owners, tracked jurisdictions, and a quarterly review cadence.

☐ I have evaluated whether a private AI environment is appropriate for the high-sensitivity HR data my team processes, and I understand the difference between public AI tools and controlled private infrastructure.

PROMPT LIBRARY & WORKBOOK EXERCISES

COMPLIANCE AUDIT PROMPT

[I want to audit our HR AI tools for compliance with [GDPR / EU AI Act / UK GDPR / CCPA]. Generate a compliance checklist specific to HR AI applications, covering: data collection and minimization, employee consent and transparency, automated decision-making obligations, data retention, and cross-border transfer requirements.]

DPIA PROMPT

[Help me conduct a simplified Data Protection Impact Assessment for an AI [recruitment / performance / L&D / employee relations] tool. Walk me through the key risk areas, the questions I need to answer for each, and the mitigations I should consider.]

WORKBOOK: The AI Inventory Exercise

List every AI tool used in your people function. For each, answer:

1. What employee data does it collect?
2. Where is that data stored?
3. Have employees been informed about it in plain language?
4. Does it make or influence individual decisions about employees? Any '*I don't know*' answers are immediate priorities.

TEAM EXERCISE: The Regulatory Horizon Scan

Assign one team member to monitor each key regulatory area: EU AI Act, GDPR, US state laws, UK ICO guidance. Meet quarterly to share updates and assess implications for your AI tools. The goal is not legal expertise, it's informed awareness. Build it into your existing team rhythm so it happens consistently, not only when something goes wrong.

SKEPTIC'S CORNER FAQ

Q: "Our legal team handles compliance. Why does HR need to understand this?"

Because HR procures, implements, and governs the tools that create the compliance risk. Legal advises you after the fact. You need enough knowledge to ask the right questions before procurement and not after a regulator makes contact. Legal cannot protect you from decisions you made before you called them.

Q: "We're a small company . . . do these regulations apply to us?"

GDPR, the EU AI Act, and most US state laws apply based on where your employees are located, not the size of your company. If you have one employee in Germany, GDPR applies to the data you hold on that employee. If you use AI in hiring decisions and have candidates in New York City, NYC Local Law 144 applies. Size is not the variable. Geography is.

Q: "The regulations are changing so fast, how do we keep up?"

You don't need to monitor everywhere. You need to monitor where it matters most, which are the jurisdictions where you have the most employees and the most AI deployment. Three to four jurisdictions, named owners, quarterly cadence. That's enough to stay ahead of most regulatory changes before they become urgent.

Q: "Our vendor told us their tool is already GDPR compliant, isn't that enough?"

No. GDPR compliance is not a certification vendors award themselves. It is a set of obligations that apply to you, **the data controller**, as well as the vendor. A vendor being compliant in how they process data does not make your procurement, deployment, or employee notification compliant. You are responsible for your end of the obligation. The vendor's claim covers theirs.

Q: "We use Microsoft Copilot. We're already in the Microsoft ecosystem. Aren't we protected?"

Being inside the Microsoft ecosystem provides stronger protections than consumer tools, but **it does not eliminate** your governance obligations. You still need to know what HR data is flowing through Copilot, whether employees have been informed, and whether your use of it would withstand a regulatory review. The ecosystem is not the governance framework. You are.

The Bottom Line

Compliance isn't a barrier to innovation. It's a signal. A signal of what responsible AI actually looks like. Transparency. Fairness. Human oversight. Accountability. These aren't regulatory burdens. They're expectations. From regulators, from employees, and from the future of work itself.

Elena's situation wasn't bad luck. It was a system failure. And systems can be redesigned.

1. Know what you're deploying.

2. Know who it affects.

3. Know what the law requires.

And know when to call a lawyer. Now you know how.

Worth Remembering

Ignorance of the law has never been a legal defense. In HR AI, it is becoming an expensive one.

> **AI Employment Laws –** *57% of HR professionals in states with enacted AI employment laws are unaware those policies exist. In New York City alone, fines for non-compliant automated hiring tools escalate at $500 to $1,500 per violation -multiplied by every day of non-compliance and every affected applicant.*
>
> ~ SHRM, State of AI in HR, 2026; HR Defense, 2025

Conclusion: The Humanly Intelligent Organization

This book does not end with a vision of the future.

It ends with a question about the present.

You have spent eleven chapters building a new literacy. You understand the zones. You understand the grey areas. You understand what AI can accelerate and where it has no business making a decision alone. You have a governance vocabulary that most of your peers don't yet have, and a framework for using it that most of your organizations haven't yet built.

The question is what you do with it before you close this book.

Figure 12.1 - Book Cover Evolution

What You Now Know That You Did Not Before

Let us be precise about what has changed.

The **Humanly Intelligence Model**

You started this book with a problem that most HR leaders are living but few can name: AI is arriving in the people function faster than the governance to contain it, faster than the literacy to evaluate it, and faster than the ethics frameworks to ensure it serves the humans it affects.

The **Humanly Intelligence Model** is the response to that problem. Not a technology strategy. Not a vendor evaluation checklist. A design principle, for every AI decision, in every HR domain, at every level of the organization.

The model holds one core claim: **AI and human judgment are not in competition**. **They are in a relationship**. And the quality of that relationship, how clearly the roles are defined, how consistently the boundaries are maintained, how deliberately the human is placed at the right point in the process, determines whether AI makes your people function better or simply faster at the wrong things.

Across recruitment, employee relations, L&D, organizational design, change management, and compliance, the model asks the same question in different contexts: where does the human belong here? Not where can the human be removed? Where must the human remain?

The Performance Diagnosis Imperative

One of the most consequential arguments in this book is one that most HR functions have never been asked to make explicit: most performance problems are not training problems.

The **Performance Analysis Quadrant** (PAQ) gave you the diagnostic infrastructure to act on that insight. Not occasionally, when a seasoned OD consultant happened to be in the room, but consistently, at scale, for every performance gap that surfaces in every team across your organization. The ghost training ecosystem that has been running quietly in your operations and sales functions for years is not a management failure. It is an infrastructure failure. Nobody gave your managers a better first question to ask.

AI changes that. By making performance consulting accessible to every HRBP, every L&D practitioner, OD Specialist and every operational manager willing to ask two questions before reaching for the course catalog.

The organizations that build this diagnostic capability now will be the ones that finally close the gap between the $400 billion invested in learning and development every year and the 74% of organizations that still cannot keep pace with their own skills demand. That gap is not a content problem. It has never been a content problem. It is a diagnosis problem. And you now have the tool to address it.

The Grey Areas Framework: A Permanent Lens

The **Grey HR Zone** is not a chapter. It is a decision-making posture that you now carry into every AI conversation you will have for the rest of your career.

The **grey area** is not the place where you do not know the answer. It is the place where the **answer depends on context, judgment, and values**, and **where that dependency is exactly the point**.

- A recruitment algorithm that produces statistically valid shortlists may still be producing them based on historical patterns that embed historical bias.
- A sentiment analysis tool that accurately identifies flight risk may still be crossing the line between people analytics and employee surveillance.
- A performance AI that reduces rating inconsistency may still be measuring the wrong things consistently.

The **grey areas framework** gives you the question behind the question.

- Not just: does this tool work? But: does it work in a way that is fair, transparent, and accountable to the humans it affects?
- Not just: is this legal? But: is it something my employees would recognize as legitimate if they could see it clearly?

That second question is the one that makes you an HR leader rather than an HR administrator.

The Agentic AI Reality

One of the most significant shifts this book asks you to make is in how you think about AI agency. Most HR leaders are still operating with a mental model of AI as a tool, something that produces an output when you give it an input, like a very fast calculator. That mental model is already obsolete.

Agentic AI systems take actions. They trigger workflows, send communications, make changes to records, and initiate processes, often without a human reviewing each step. They are already in your HR function. They are in your HRIS, your recruitment platform, your onboarding workflow, your compliance notification system. You may not have procured them as agentic systems. That does not change what they are doing.

Knowing where agentic systems operate, what decisions they make, and where human checkpoints must exist is not just good governance. It is a competitive advantage. Most organizations wait until something goes wrong. You now have the framework to act before it does.

The mental model that holds: the AI agent acts as the engine. The human remains the steering wheel. Power without direction is not an asset. It is a liability.

The Ethical Foundations: Professional Identity, Not Just Compliance

The compliance chapters in this book exist **because the regulatory environment is real**, **consequential**, and **moving faster** than most organizations' awareness of it. But compliance is the floor, not the ceiling.

The deeper argument, the one this book has been building toward from Chapter 1, is about professional identity. HR exists to ensure that the people who come to work for an organization are treated with fairness, dignity, and transparency. That is not a mission statement. It is a professional obligation. And it does not get suspended because a vendor's algorithm is faster than a human reviewer.

Every AI tool that makes or influences a decision about a person's employment is carrying part of that obligation. If it is biased, **HR is responsible**. If it is opaque, **HR is responsible**. If it operates without meaningful human oversight, **HR is responsible**. Not because HR procured it, though often HR did, but because HR is the function whose professional identity is bound to the fair treatment of employees.

Getting this right is not compliance. It is the definition of the profession in the age of AI.

The Three Commitments

Knowledge without commitment is just awareness. Awareness is not enough. Before you close this book, there are three commitments worth making, to yourself, to your team, and to the employees whose working lives your people function shapes.

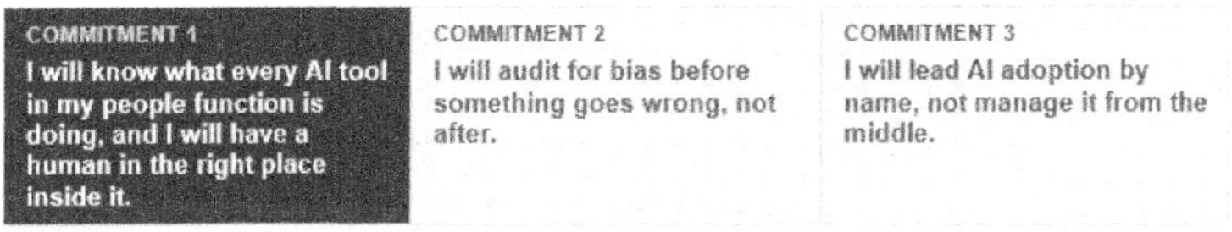

Figure 12.2 - Three Commitments

Commitment 1: I Will Know What Every AI Tool in My People Function Is Doing

This is the foundational commitment, and the one most likely to reveal uncomfortable gaps. Inventory is unglamorous work. It does not appear in leadership competency frameworks or performance reviews. But an HR leader who cannot answer the question "what employee data are our AI tools collecting right now?" is operating without the basic situational awareness that governance requires.

Know your tools. Know what they collect. Know what they decide. Know where the human is, and where the human is absent. Then fix the absence.

Having a human in the right place inside it is not a passive commitment. It requires actively designing human checkpoints into workflows that were built without them and actively pushing back on vendors and internal technology teams who treat human review as friction to be minimized. It requires making the case, and sometimes repeatedly, that the goal is not the fastest workflow. The goal is the right decision.

You are not just a user of tools. You are the architect of their guardrails.

Commitment 2: I Will Audit for Bias Before Something Goes Wrong

Bias auditing is the commitment that gets deferred most often. Not because HR leaders do not understand the importance, they do, but because it is the work that has no immediate deadline, no visible output, and no organizational pressure until the bias surfaces publicly and the deadline becomes a crisis.

The commitment to audit proactively is a commitment to a different relationship with risk. Not risk aversion. Risk intelligence. Understanding where the bias risk is highest in your AI-assisted processes, designing an audit process that is rigorous enough to find it before it harms someone, and building the organizational relationship with that finding that allows you to act on it without defensiveness.

This is also where the Grey HR Zone is most consequential. Bias in AI-assisted recruitment, performance assessment, or promotion decisions is not a technical failure. It is a fairness failure. And fairness failures in people decisions are the failures that destroy trust, with employees, with candidates, and with the regulators who are increasingly watching.

Proactive ethics is the only ethics that counts in the AI era. Demand transparency from your vendors. Conduct red team exercises within your own team to find the cracks in your logic before they become chasms in your culture.

Commitment 3: I Will Lead AI Adoption by Name

The third commitment is the one about visibility. Managing AI adoption from the middle, implementing tools, briefing teams, solving problems as they arise, but never explicitly naming the governance framework you are building or presenting it as a standard, is a form of professional underreach that the moment does not permit.

HR leaders who lead AI adoption by name, who stand in front of their executive teams and say "this is our AI governance framework, this is the standard we are setting, and this is why it matters for our people and our organization," are the ones who shape organizational culture around AI, not just respond to it.

Leading by name means claiming the authority that the professional obligation gives you. It means saying plainly: the people function sets the standard for how AI treats people in this organization, because that is what the people function is for. It means owning the narrative, ensuring that your employees hear about AI's role in their future from the people function first, and with a focus on empowerment, not replacement. It means making the work visible enough that it can be held to account, and visible enough that it can be replicated, funded, and built upon.

Manage from the middle and you will always be cleaning up after decisions that were made without you.

Lead by name and you get to shape the decisions before they are made.

Your 90-Day Action Plan

Three commitments without a plan are intentions. Here is the plan.

Phase	Focus	The Work
Days 1-30	The Audit	☐ Complete the Humanly Intelligence Zone Sorter for every active AI workflow in your people function. ☐ Map out what you have. ☐ Classify what it does. ☐ Identify every place where a human checkpoint is missing. Do not fix anything yet. See the full picture first.
Days 31-60	The Fix	☐ Implement human checkpoints for every Zone 2 and Zone 3 workflow that does not have one. ☐ Update your AI vendor contracts for the compliance clauses in Chapter 10. ☐ Brief your team on the Grey HR Zone framework. ☐ Run the PAQ diagnostic on your three most recent training requests and confirm training was the right intervention for each. ☐ Run your first bias audit on your highest-volume AI-assisted decision.
Days 61-90	The Lead	☐ Present your AI governance framework to your executive team. Name the standard you are setting. ☐ Propose the Humanly Intelligence principles as organizational policy, not just HR practice. ☐ Set the review cadence. ☐ Make the work visible, because visible governance is governance that holds.

Figure 12.3 - 90-Day Action Plan

A Note on the 90-Day Frame

Ninety days is not arbitrary. It is the window between **reading something** and either acting on it or **filing it** in the category of things you meant to do. It is also the window within which a new framework either becomes embedded in organizational practice or gets displaced by the next priority.

> The three phases are designed to build on each other. You cannot fix what you have not audited. You cannot lead what you have not fixed. The sequence is the discipline.

The appendices in this book are the tools for each phase. The Zone Sorter for the audit. The Vendor Vetting Matrix and Agentic Workflow Registry for the fix. The HR AI Ethical Charter for the lead. They are not decorative back matter. They are the implementation infrastructure for the three commitments.

A WORD ON MOMENTUM

The 90-day plan works because it starts with something you can do alone, in your own function, with no organizational permission required.

The audit is yours to run. The human checkpoints are yours to implement. You do not need executive sponsorship to map your AI workflows or to add a review step to the Zone 2 process.

By the time you reach Day 61, you will have something to show. A completed inventory. A governance framework with evidence behind it. A bias audit with findings. That is the material that earns the executive conversation, not a request for permission to start, but a report on what you have already built.

Don't wait for the organization to ask for this. Build it. Then show them what it looks like.

Figure 12.4 - Momentum

A Final Word on What Makes HR Human

Here is the argument this book has been building toward.

AI does not threaten the HR profession.

The absence of human judgment in AI-driven people work does.

That distinction matters because it changes what HR leaders are being asked to do. If the threat is AI itself, the response is resistance, slow the adoption, limit the tools, and insist on human processes because they are human. That response is already losing. The adoption is happening. The tools are arriving. Resistance is not a strategy.

If the threat is the absence of human judgment, the response is something different entirely. It is presence. It is deliberate, informed, structurally embedded human judgment at every point in the AI-assisted process where it matters. Not judgment as a rubber stamp. Judgment as the thing that makes the process legitimate.

This is a better fight. It is a fight HR is equipped for. And it is the fight that determines whether the HR profession emerges from the AI era with more influence than it entered with, or less.

The HR Leader Who Thrives in the AI Era

- It is not the HR leader who uses AI the most.
- It is not the HR leader who has automated the most workflows, deployed the most tools, or achieved the highest efficiency metrics from AI-assisted processes.
- It is the HR leader who knows exactly where not to.

The one who looks at an automated performance rating system and says: not without a manager conversation. The one who looks at an AI-generated shortlist and says: not without an audit. The one who looks at an employee sentiment tool and says: not without employee transparency. The one who holds the line, and not because they don't understand the technology, but because they understand it well enough to know where it ends and where the human must begin.

Because while a machine can calculate the probability of success for a candidate, only a human can feel the spark of potential that defies the data.

That knowledge is not a limitation. It is the most sophisticated thing an HR leader can bring to an AI-saturated organization. It is the capability that cannot be automated, replicated, or replaced, because it requires professional judgment, ethical grounding, and the willingness to say no to speed when speed is the wrong answer.

Humanly Intelligence

Humanly Intelligence is not a compromise between AI and humanity.

It is not a slowing down of AI to make room for humans. It is not a softening of efficiency in the name of values. It is not a cautious middle ground between the technology optimists and the technology critics.

A core principle: when AI and human judgment work together, outcomes are more consistent, more legitimate, and more accountable.

The organizations that build on this principle will make better hiring decisions. They will develop their people more effectively. They will manage employee relations with more fairness and less liability. They will adopt AI in ways their employees trust rather than resent. And they will be better positioned to navigate whatever comes next, because they will have built the governance infrastructure to evaluate it clearly rather than react to it under pressure.

That is the Humanly Intelligent Organization.

That is what this book has been building toward.

THE QUESTION BEFORE YOU CLOSE THIS BOOK

Not: what will AI do to HR?

That question belongs to the future, and the future is genuinely uncertain.

The question is: What will you do in the next 90 days to ensure AI in your organization serves people with fairness, transparency, and accountability?

That question has an answer.

You now know how to find it.

Figure 12.5 - Important Question

Look at your hands.

They are the hands of a leader in the most important era in the history of work. The tools are ready. The framework is in your mind.

Now go build the **Humanly Intelligent Organization**.

COMPANION RESOURCES

All templates, checklists, and prompt libraries in this book are available as downloadable resources at the Shelf-Help Press | Humanly Intelligence for HR Leaders resource site. Google Drive link: https://drive.google.com/drive/folders/1cT_4BkQzpifVEElITVzaTV1-cS85UXQ7?usp=sharingOR Select the QR Code below to access.

For Audible listeners, send an email to Renee@3KARATproductions.com and in the subject area type the word **Grey HR Zone.**

HUMANLY INTELLIGENCE for HR LEADERS MATERIALS

Resources are updated regularly as AI capabilities and regulatory requirements evolve.

A facilitation guide for using this book as a team learning program is also available for download.

Join the Humanly Collective community at https://www.linkedin.com/humanly-collective/humanly-collective to connect with other people professionals navigating AI adoption.

References & Further Reading

References & Further Reading

American Bar Association. (2024). Navigating the AI employment bias maze: Legal compliance guidelines and strategies. Business Law Today. https://www.americanbar.org/groups/business_law/resources/business-law-today/2024-april/navigating-ai-employment-bias-maze/

American Hospital Association. (2020). Workforce case study: Artificial intelligence automating processes for professionals. AHA Center for Health Innovation. https://www.aha.org/system/files/media/file/2020/02/workforce-case-study-artificial-intelligence-automating-processes-for-professionals-2-2020.pdf

Barocas, S., & Selbst, A. D. (2016). Big data's disparate impact. California Law Review, 104(3), 671–732.

Bersin, J. (2023, April 9). How AI is disrupting the HR tech marketplace. Josh Bersin. https://joshbersin.com/2023/04/how-ai-is-disrupting-the-hr-tech-marketplace/

Bersin, J. (2023, August 18). Welcome to the post-industrial economy: AI is here to save us. Josh Bersin. https://joshbersin.com/2023/08/welcome-to-the-post-industrial-economy-ai-is-here-to-save-us/

Bersin, J. (2026, February 11). New research: How AI transforms $400 billion of corporate learning. Josh Bersin. https://joshbersin.com/2026/02/new-research-how-ai-transforms-400-billion-of-corporate-learning/

Bersin, J. (2026, March 26). Why AI is a massive job-creation technology, despite what you think. Josh Bersin. https://joshbersin.com/2026/03/why-ai-is-a-massive-job-creation-technology-despite-what-you-think/

Bersin, J. (2026, April 6). Introducing HR 2030: A vision for agentic human resources. Josh Bersin. https://joshbersin.com/2026/04/introducing-hr-2030-a-vision-for-agentic-human-resources/

Bursell, M., & Roumbanis, L. (2024). After the algorithms: A study of meta-algorithmic judgments and diversity in the hiring process. Big Data & Society, 11(1). https://journals.sagepub.com/doi/10.1177/20539517231221758

Christian, B. (2020). The alignment problem: Machine learning and human values. W. W. Norton.

Crawford, K. (2021). Atlas of AI: Power, politics, and the planetary costs of artificial intelligence. Yale University Press.

Crowell & Moring. (2026). Artificial intelligence and human resources in the EU: A 2026 legal overview. https://www.crowell.com/en/insights/client-alerts/artificial-intelligence-and-human-resources-in-the-eu-a-2026-legal-overview

Dahl, M. J., Muehlematter, V. V., Nie, J. Z., & Ho, D. E. (2024). Large language models on trial: Legal models hallucinate in 1 out of 6 (or more) benchmarking queries. Stanford Institute for Human-Centered AI (HAI).

Dastin, J. (2018, October 10). Amazon scraps secret AI recruiting tool that showed bias against women. Reuters. https://www.reuters.com/article/us-amazon-com-jobs-automation-insight/amazon-scraps-secret-ai-recruiting-tool-that-showed-bias-against-women-idUSKCN1MK08G

Daugherty, P., & Wilson, H. J. (2018). Humans + machines: Reimagining work in the age of AI. Harvard Business Review Press.

Davenport, T. H., & Harris, J. G. (2017). Competing on analytics: The new science of winning (updated ed.). Harvard Business Review Press.

Deloitte. (2023a). Global human capital trends. Deloitte Insights.

Deloitte. (2023b). NYC Local Law 144-21 and algorithmic bias. https://www.deloitte.com/us/en/services/audit-assurance/articles/nyc-local-law-144-algorithmic-bias.html

Edmondson, A. C. (2018). The fearless organization: Creating psychological safety in the workplace for learning, innovation, and growth. Wiley.

Epstein Becker Green. (2023). Taking stock of New York City's automated employment decision tools law. Workforce Bulletin. https://www.workforcebulletin.com/taking-stock-of-new-york-citys-automated-employment-decision-tools-law

Equal Employment Opportunity Commission. (1978). Uniform guidelines on employee selection procedures. https://www.eeoc.gov/laws/guidance/uniform-guidelines-employee-selection-procedures-0

European Parliament and Council of the European Union. (2024). Regulation (EU) 2024/1689 laying down harmonised rules on artificial intelligence (Artificial Intelligence Act). Official Journal of the European Union. https://digital-strategy.ec.europa.eu/en/policies/regulatory-framework-ai

Fuller, J. B., Raman, M., Sage-Gavin, E., & Hines, K. (2021). Hidden workers: Untold talent. Harvard Business School Project on Managing the Future of Work and Accenture. https://www.hbs.edu/managing-the-future-of-work/Documents/research/hiddenworkers090321.pdf

Gartner. (2023). HR technology market guide. Gartner Research.

Gilbert, T. F. (1978). Human competence: Engineering worthy performance. McGraw-Hill.

Guenole, N., Ferrar, J., & Feinzig, S. (2017). The power of people: Learn how successful organizations use workforce analytics to improve business performance. FT Press.

Hastings, R., & Meyer, E. (2020). No rules rules: Netflix and the culture of reinvention. Penguin Press.

Hiatt, J. (2006). ADKAR: A model for change in business, government and our community. Prosci Learning Center.

Highhouse, S. (2008). Stubborn reliance on intuition and subjectivity in employee selection. Industrial and Organizational Psychology, 1(3), 333–342. https://doi.org/10.1111/j.1754-9434.2008.00058.x

Hunton Andrews Kurth. (2024). The impact of the EU AI Act on human resources activities.

https://www.hunton.com/insights/legal/the-impact-of-the-eu-ai-act-on-human-resources-activities

IBM. (2024). The rise of AI agents: From assistants to coworkers. IBM Think. https://www.ibm.com/think/topics/ai-agents

Jobin, A., Ienca, M., & Vayena, E. (2019). The global landscape of AI ethics guidelines. Nature Machine Intelligence, 1(9), 389–399.

Kaur, P., & Sahdev, S. (2024). The impact of generative AI on recruitment: Navigating the era of AI-generated resumes. Journal of Applied Human Resources Management, 5(2), 112–128.

Kelan, E. K. (2023). Algorithmic inclusion: Shaping the predictive algorithms of artificial intelligence in hiring. Human Resource Management Journal, 33(4). https://onlinelibrary.wiley.com/doi/10.1111/1748-8583.12511

Koenecke, A., Nam, A., Lake, E., Nudell, J., Quartey, M., Mengesha, Z., & Goel, S. (2020). Racial disparities in automated speech recognition. Proceedings of the National Academy of Sciences, 117(14), 7684–7689.

Kotter, J. P. (2012). Leading change (updated ed.). Harvard Business Review Press.

Littler Mendelson. (2023). EEOC issues guidance on use of artificial intelligence tools in employment selection procedures under Title VII. https://www.littler.com/publication-press/publication/eeoc-issues-guidance-use-artificial-intelligence-tools-employment

Maurer, R. (2024). Hiring managers are struggling with AI-generated candidate spam. Society for Human Resource Management. https://www.shrm.org/topics-tools/news/talent-acquisition/hiring-managers-struggling-ai-generated-candidate-spam

Mayer Brown. (2023). EEOC issues Title VII guidance on employer use of AI and other algorithmic decision-making tools. https://www.mayerbrown.com/en/insights/publications/2023/07/eeoc-issues-title-vii-guidance-on-employer-use-of-ai-other-algorithmic-decisionmaking-tools

Mesriani, K. (2024). AI & HR: Algorithmic discrimination in the workplace. Cornell Journal of Law and Public Policy. https://publications.lawschool.cornell.edu/jlpp/2024/11/21/ai-hr-algorithmic-discrimination-in-the-workplace/

Mollick, E. R. (2024). Co-intelligence: Living and working with AI. Portfolio/Penguin.

National Institute of Standards and Technology. (2023). Artificial Intelligence Risk Management Framework (AI RMF 1.0). U.S. Department of Commerce. https://doi.org/10.6028/NIST.AI.100-1

New York City Council. (2021). Local Law 144 of 2021: Automated employment decision tools. https://www.nyc.gov/site/dca/about/automated-employment-decision-tools.page

Northwestern Journal of Technology and Intellectual Property. (2025). Algorithmic bias in AI employment decisions. https://jtip.law.northwestern.edu/2025/01/30/algorithmic-bias-in-ai-employment-decisions/

O'Neil, C. (2016). Weapons of math destruction: How big data increases inequality and threatens democracy. Crown.

Ogletree Deakins. (2025). The EU AI Act is here — what it means for US employers. https://ogletree.com/insights-resources/blog-posts/cybersecurity-awareness-month-in-focus-part-iii-the-eu-ai-act-is-here-what-it-means-for-u-s-employers/

Pena, A., et al. (2023). Ethics and discrimination in artificial intelligence-enabled recruitment practices. Humanities and Social Sciences Communications, 10, Article 567. https://www.nature.com/articles/s41599-023-02079-x

Prince, J., & Schwarcz, D. (2020). Proxy discrimination in the age of algorithms and big data. Iowa Law Review, 105(3), 1257–1318.

Raghavan, M., Barocas, S., Kleinberg, J., & Levy, K. (2020). Mitigating bias in algorithmic hiring: Evaluating claims and practices. Proceedings of the 2020 Conference on Fairness, Accountability, and Transparency, 469–481. https://doi.org/10.1145/3351095.3372828

Rogers, E. M. (2003). Diffusion of innovations (5th ed.). Free Press.

Rummler, G. A., & Brache, A. P. (1990). Improving performance: How to manage the white space on the organization chart. Jossey-Bass.

Schneier, B. (2015). Data and Goliath: The hidden battles to collect your data and control your world. W. W. Norton.

ScienceDirect. (2025). Bias in AI-driven HRM systems: Investigating discrimination risks embedded in AI recruitment tools and HR analytics. https://www.sciencedirect.com/science/article/pii/S2590291125008113

Senge, P. M. (1990). The fifth discipline: The art and practice of the learning organization. Doubleday.

Society for Human Resource Management. (2023). SHRM benchmarking reports: Talent acquisition and retention. SHRM.

Steiner, K. (2017, March 15). Bad candidate experience cost Virgin Media $5M annually — here is how they turned that around. LinkedIn Talent Solutions. https://business.linkedin.com/talent-solutions/blog/candidate-experience/2017/bad-candidate-experience-cost-virgin-media-5m-annually-and-how-they-turned-that-around

Tandfonline. (2025). Reducing AI bias in recruitment and selection: An integrative grounded approach. International Journal of Human Resource Management. https://www.tandfonline.com/doi/full/10.1080/09585192.2025.2480617

The Talent Board. (2024). 2023 candidate experience (CandE) global benchmark research report. https://www.talentboard.org/benchmark-research/cande-reports/

U.S. Equal Employment Opportunity Commission. (2023, May 18). Assessing adverse impact in software, algorithms, and artificial intelligence used in employment selection procedures under Title VII of the Civil Rights Act of 1964 [Technical assistance document]. https://www.eeoc.gov/laws/guidance/questions-and-answers-clarify-and-provide-common-interpretation-uniform-guidelines

Ulrich, D., Younger, J., Brockbank, W., & Ulrich, M. (2012). HR from the outside in: Six competencies for the future of human resources. McGraw-Hill.

Wilson, K., & Caliskan, A. (2024). AI tools show biases in ranking job applicants' names according to perceived race and gender. University of Washington. Presented at AAAI/ACM Conference on AI, Ethics, and Society.

https://www.washington.edu/news/2024/10/31/ai-bias-resume-screening-race-gender/

Workday. (2025). The AI-enabled workforce: Applications and benefits of AI in HR. Workday Newsroom.

Printed in Dunstable, United Kingdom

91232333R00150